Mammalian Tumor Cell Heterogeneity

Authors

John T. Leith, Ph.D.
Professor of Medical Sciences
Division of Biology and Medicine
Brown University
and
Department of Radiation Oncology
Rhode Island Hospital
Providence, Rhode Island

Daniel L. Dexter, Ph.D.
Research Associate
Biomedical Products Department
E.I. DuPont De Nemours and Company
Wilmington, Delaware
and
Adjunct Associate Professor of Medicine
Brown Univeristy
Providence, Rhode Island

CRC Press, Inc.
Boca Raton, Florida

Library of Congress Cataloging in Publication Data

Leith, John T.
  Mammalian tumor cell heterogeneity.

  Includes bibliographies and index.
1. Tumors.   2. Cancer cells.   3. Mammals--
Diseases.   I. Dexter, Daniel L., 1940-
[DNLM: 1. Cell Transformation, Neoplastic.
2. Neoplasms--etiology.   3. Neoplasms--therapy.
QZ 202 L533m]
RC267.L44   1986   599′.02   85-7855
ISBN 0-8493-6162-1

  This book represents information obtained from authentic and highly regarded sources. Reprinted material is quoted with permission, and sources are indicated. A wide variety of references are listed. Every reasonable effort has been made to give reliable data and information, but the author and the publisher cannot assume responsibility for the validity of all materials or for the consequences of their use.

  All rights reserved. This book, or any parts thereof, may not be reproduced in any form without written consent from the publisher.

  Direct all inquiries to CRC Press, Inc., 2000 Corporate Blvd., N.W., Boca Raton, Florida, 33431.

© 1986 by CRC Press, Inc.

International Standard Book Number 0-8493-6162

Library of Congress Card Number 85-7855
Printed in the United States

PREFACE

This text discusses aspects of cellular and environmental diversity that exist within individual cancers which may be of human or nonhuman origin, and may be either primary tumors or metastatic deposits from primary neoplasms. It is now quite evident that the phenomenon of intratumor heterogeneity exists. While many investigators perceive this heterogeneity to mean only that multiple subpopulations of tumor cells that express different phenotypes are present in solid tumors, it must not be forgotten that heterogeneity is not just a static differential expression of a particular phenotype, but also includes the dynamic aspects of cellular proliferation, and modulation of proliferation and phenotypic expression by the local tumor environment. Such diversity presents the experimenter and the clinician with hard problems in regard to the study and treatment of cancer. Questions such as how does one accurately define and describe the extent of heterogeneity, or the problem of the optimal strategy of therapy loom large. The presence of such intra- and interlesional heterogeneity appears to be a mechanism for the survival of the tumor, and efforts to combat such a difficult opponent are under active pursuit at present. While this text reviews the field of tumor heterogeneity and has as its impetus the renewed interest in this area that has arisen in the past decade, it is relevant to note that documentation of human tumor heterogeneity as indicated by histology had been provided by Virchow over a hundred years ago. However, the importance of this diversity was not appreciated by the great majority of the scientific and medical community, and indeed the concept of cancer was that of a disease that was essentially unchanging, albeit outside of normal cellular growth control mechanisms. However, this concept is now appreciated to be incomplete, and with the introduction of new heterogeneous tumor models, of both rodent and human origin, much clever investigation has been done into the implications and ramifications of intraneoplastic diversity. A central feature of the work in tumor heterogeneity has been the multidisciplinary nature of the studies that have provided insight, from the basic morphological description of heterogeneity to biochemical, therapeutic, etc. studies that have all been necessary to provide a global appreciation of the phenomenon.

We have intended this volume to be a current review of the status of tumor heterogeneity. It has provided the authors with a chance to coalesce some of their thoughts in this area, and we hope that this work will be of use to graduate and postdoctoral students in the field of tumor biology and closely related disciplines, and will provide other investigators in the field of mammalian tumor cell heterogeneity with a rounded view. We have tried to provide a relatively inclusive set of references for each area of tumor heterogeneity that is discussed. The general organization of this volume has been to proceed from the historical basis of tumor heterogeneity, to the evolution of thought with regard to the *raison d'être* for the existence of the phenomenon, to other areas that are required for an overall appreciation of tumor heterogeneity (e.g., cell kinetic aspects and environmental modulation of tumor heterogeneity). We have discussed the therapeutic aspects, and have tried to point out the major problems for treatment associated with the expression of the metastatic phenotype of heterogeneous tumors. Clearly, some areas of this volume are speculative, as little actual data exists for some of the aspects of tumor heterogeneity. In these areas, we have attempted to indicate where additional research efforts would be of great help in the understanding of intraneoplastic diversity. In many instances, the questions raised greatly outnumber the answers provided. We hope, however, that this gives the reader, as it does us, a sense of excitement with respect to the possibilities of new insights into basic aspects of tumor biology and their clinical implications.

John T. Leith
Daniel L. Dexter

The authors would like to dedicate this volume to the memory of Dr. Charles Heidelberger.

THE AUTHORS

John T. Leith, Ph.D., is currently Chief Radiobiologist in the Department of Radiation Medicine, Rhode Island Hopsital, Providence, Rhode Island. He is also Professor of Medical Sciences in the Division of Biology and Medicine at Brown University in Providence. His teaching activities include courses in radiation biology and biophysics at both the undergraduate and graduate levels. General research interests lie in the area of the effects of ionizing radiation and hyperthermia in mammalian cells, both normal and malignant. Current research interests include studies of the responses of heterogeneous tumor systems to cytotoxic agents, and are intended to provide basic information both in tumor biology and in possible strategic approaches to tumor therapy.

Dr. Leith received his Ph.D. degree at Boston University where he was a NASA fellow. He was an AEC Postdoctoral fellow at the Lawrence Berkeley Laboratory, University of California, Berkeley, and has been a staff research biophysicist at the Lawrence Berkeley Laboratory, and later, Associate Professor at the University of Arizona, Tucson, prior to his current position in Rhode Island. Dr. Leith is an active member of the Radiation Research Society and the American Association for Cancer Research.

Daniel L. Dexter, Ph.D., is currently Research Associate in the Cancer Chemotherapy Program, Pharmaceuticals Division of the Biomedical Products Department, E.I. DuPont De Nemours & Co., Inc., Wilmington, Delaware. He also holds the appointment of Adjunct Associate Professor of Medicine at Brown University, Providence, Rhode Island. His research interests are in the areas of cancer cell differentiation, heterogeneity, and therapeutics.

Dr. Dexter received his Ph.D. degree in Oncology from the McArdle Laboratory of the University of Wisconsin, Madison. His major professor was Dr. Charles Heidelberger. Dr. Dexter received a stipend from the Jane Coffin Childs Foundation to study as a postdoctoral fellow at the Pasteur Institute, Paris, under the direction of Dr. Francois Jacob. Dr. Dexter has held the positions of Instructor, Assistant Professor, and Associate Professor of Medicine at Brown University prior to accepting his current position with the DuPont Company.

ACKNOWLEDGMENTS

The authors would like to express their appreciation to Dr. Stephen Baylin of the Johns Hopkins University, Baltimore, Maryland and Dr. Daniel Von Hoff of the University of Texas Medical School, San Antonio, Texas for helpful comments and use of unpublished material in this text. Also, we would like to thank Ms. Tammi McHugh, Ms. Sara Lafreniere, and Ms. Christine Gailey for their help in the preparation of this text.

TABLE OF CONTENTS

Chapter 1
Introduction to Mammalian Tumor Cell Heterogeneity 1
I. Introduction 1
II. The Paradox of Murine Tumor Models 2
III. Heterogeneity Within Murine Tumors 3
IV. Phenomenology of Murine Tumor Heterogeneity 3
    A. Immunological Heterogeneity 3
    B. Heterogeneity for Chromosomal and Biochemical Markers 4
    C. Heterogeneity for Invasive and Metastatic Potential 6
    D. Heterogeneity for Response to Treatment Modalities 7
V. Summary 7
References 7

Chapter 2
Intraneoplastic Diversity Within Human Tumors
I. Introduction 11
II. Intratumor Heterogeneity Demonstrated in Patient Material 11
    A. Breast Carcinoma 11
    B. Lung Cancer 13
    C. Colorectal Cancer 15
    D. Neuroblastoma 16
    E. Medullary Thyroid Carcinoma 16
    F. Other Tumors 17
        1. Ovarian Tumors 17
        2. Brain Tumors 18
        3. Other Examples 18
III. Summary 19
References 19

Chapter 3
The Nowell Hypothesis and Subsequent Developments
I. Introduction 23
II. The Dynamic Tumor 25
III. Maturational Therapy of Heterogeneous Tumors 29
IV. Tumor Heterogeneity and Oncogenes 30
V. Summary 31
References 32

Chapter 4
Metastasis
I. Introduction 35
II. The Metastatic Process 36
III. Metastasis — A Random or Selective Process? 37
IV. Are Metastases Clonal? 39
V. Summary 41
References 41

Chapter 5
Quantitative Aspects of Tumor Heterogeneity
I. Introduction..45
II. Analytical Aspects of Tumor Heterogeneity 45
    A. Dissociation and Cell Separation Procedures 45
    B. Centrifugal Elutriation in the Study of Heterogeneous Systems.........48
    C. Other Cell Separation Procedures 50
    D. Flow Cytometric (FCM) Techniques 51
    E. Cell Cycle Consideration in Heterogeneous Tumors 56
III. Clinical Evidence of Kinetic Tumor Heterogeneity 57
    A. Studies of Various Tumor Types..58
    B. Lung Tumors (Nonsmall Cell Carcinomas)..................................60
    C. Lung Tumors (Small Cell Carcinomas).......................................62
IV. Summary..62
References 64

Chapter 6
Environmental Aspects of Tumor Heterogeneity
I. Introduction..69
II. Tumor Hypoxia 69
III. Tumor pH 71
IV. Nutrients...72
V. Other..74
VI. Summary and Conclusions 74
References 75

Chapter 7
Clonal Interactions Among Tumor Subpopulations
I. Introduction..79
II. Requirements for Observation of Clonal Interactions 79
III. In Vivo Experimental Studies 79
IV. In Vitro Experimental Studies 82
V. In Vivo Metastatic Assay Systems 85
VI. Tumor - Host Cell Interactions...91
VII. Summary..92
References 93

Chapter 8
Responses of Heterogeneous Primary Tumors to Therapy
I. Introduction..97
II. Responses to Chemotherapeutic Agents...97
III. Responses to Ionizing Radiation 100
IV. Responses to Hyperthermia 104
V. Responses to Combined Modality Treatments................................... 110
VI. Relative Ranking of Responses to Various Cytotoxic Agents 114
VII. Summary.. 115
    A. Distribution of Responsivity within Heterogeneous Tumors 115
    B. Selection of Resistant Subpopulations 116
    C. Random Association of Response Phenotypes 116
    D. Clinical Approach to Heterogeneous Tumors............................. 117
References 117

Chapter 9
Therapy of Metastatic Cancers
I. Introduction 121
II. Experimental Models of Metastasis 121
III. Laboratory Studies on the Sensitivity of Metastatic Cells to Chemotherapeutic
    Agents 122
IV. Clinical Studies 124
V. Approaches to the Therapy of Metastatic Disease 124
VI. Commonalities in Metastases 126
    A. Ploidy 126
    B. Cytoskeleton 127
    C. Tumor Angiogenesis Factor 128
    D. Tumoricidal Macrophages 128
    E. Cancer Cell Differentiation 128
    F. Tumor Cell Kinetics 129
    G. Oncogenes and their Products 129
VII. Commonalities from Another Perspective 130
References 131

Chapter 10
Summary and Conclusions 137

Index 139

Chapter 1

INTRODUCTION TO MAMMALIAN TUMOR CELL HETEROGENEITY

"Analysis of the morphology of mammary tumors disposes forever of the idea that cancer is usually a unit alteration in a single cell, which reproduces itself unchangeably." Thelma Dunn, 1959.[1]

I. INTRODUCTION

Diversity has been the hallmark of life on our planet. Evolution has resulted in the appearance of increasingly complex creatures on earth. These living beings are each composed of organ systems that are functionally different from one another. The component organs of higher life forms are in turn comprised of specialized tissues and subsets of cells. This diversity, which characterizes all of our natural world, also characterizes a pathological life form uniquely equipped to grow and survive: the malignant neoplasm.

This volume will cover various aspects of diversity within individual cancers. The clinical importance of intratumor heterogeneity depends on the extent to which intraneoplastic diversity occurs within human cancers. It is now certain that the phenomenon exists in most, if not all, human solid tumors. The implications for such diversity within human neoplasms are apparent almost at a first glance, and preclinical and clinical investigations have amply confirmed that tumor heterogeneity has profound ramifications for the management of cancer patients. Ominously, the variability within single neoplasms has a direct and causal relationship to the dissemination of these tumors, an event which is responsible for the majority of cancer deaths. Progress in understanding cancer, including the propensity of cancer cells to colonize distant sites, and in designing effective therapy to treat the disease will depend to a large extent on our ability to unravel the tangled knot presented by tumor heterogeneity. Diversity within neoplasms is essential to the tumor's ability to survive, grow, and metastasize, and it is perhaps the major problem facing experimental and medical oncologists.

One could argue that a direct examination of tumor heterogeneity in human cancers is the obvious place to begin a discussion, since it is the human disease that concerns us. Indeed, variability within human neoplasms has been recognized by clinical pathologists for a very long time now.[2] Investigators caught up in the resurgence of interest in tumor heterogeneity are often surprised to find that diversity in human cancers had been documented by Virchow more than one hundred years ago.[3] Bigner et al.,[4] in a recent review article, have provided an excellent history of reports of heterogeneity in human gliomas from Virchow until the present time. Thirty years after Virchow's description of heterogeneity in glioblastoma multiforme (hence the name), Stroebe, in 1895, provided a classic definition of a heterogeneous human neoplasm, the gliosarcoma.[5] The translation by Bigner et al. of Stroebe's original report in German shows clearly that mixed tumors containing glial and sarcomatous elements might be expected to occur. It is fascinating that Bigner and colleagues presented evidence of just such a tumor using an immunohistochemical assay for glial fibrillary acidic protein (GFAP). They demonstrated that the particular gliosarcoma they studied indeed contained nests of GFAP-positive tumor cells surrounded by GFAP-negative cancer cells.[4]

Other workers early in this century also documented the cellular heterogeneity that characterizes brain tumors.[6-9] The conclusion drawn by the Duke authors is that glioblastomas have not changed in their appearance (heterogeneity) in more than 100 years. Evidence presented in Chapters 2 and 5 convincingly demonstrates that heterogeneity in human solid tumors can be found whenever one looks for it. It is probably a safe conclusion that such heterogeneity has existed in these neoplasms for as long as cancer has afflicted mankind.

However, these early observations on human tumors were, in general, lost on the great majority of the scientific and medical community. For this reason, studies with murine tumors have been extremely important in stimulating research on the occurrence and importance of diversity within neoplasms.

II. THE PARADOX OF MURINE TUMOR MODELS

Paradoxically, work with mouse tumors has pushed our thinking about heterogeneity along two diametrically opposite paths. On the one hand, work done with murine mammary carcinomas many decades ago has provided us with strong experimental evidence that led to the conclusion that tumors are heterogeneous tissues in the sense that they are each comprised of distinct subpopulations of neoplastic cells.[1,10] Indeed, where tumor heterogeneity was appreciated early as a phenomenon of intrinsic, cellular tumor diversity, it was generally so perceived by those studying murine tumors. Clinical pathologists often attributed histological variability within human cancers, or between a primary tumor and its metastases, as being due to host or environmental factors modifying the tumor's appearance. The belief that the diversity within tumors might be due to heterogeneity within the tumor itself, i.e., due to the existence of distinct neoplastic subpopulations, was limited mainly to a few scientists working with mouse tumors and to a handful of insightful physicians. This idea was eloquently expressed by Thelma Dunn over 25 years ago, as can be seen from the quotation given at the beginning of this chapter.[1]

However, while a few workers were contemplating the idea of a tumor consisting of clones of cells, changing with time, and progressing to a more malignant state, many other investigators were thinking just the opposite. The prevailing belief until the past 5 to 10 years was that tumors were monolithic, unchanging, pathologically aberrant growths, consisting of tumor cells that divided out of control and bred true.[11,12] The neoplasm was composed of a homogeneous collection of replicate, identical cells. Metastasis resulted from the colonization of distant sites by a few of these cells. The dissemination process could be dependent on purely mechanical factors,[13] or might depend on the "soil" of the target organ.[14] However, the process would be random with respect to the cancer cells themselves; every cell would have an equal, albeit low, probability of metastasizing. Any variability observed within a mouse (or human) neoplasm, or between a primary and secondary tumor, would be due to the "soil" of the involved host tissue, or to immunological factors, or to the differential availability of oxygen, nutrients, and effective waste disposal between different regions of a tumor. Diversity would not be due to intrinsic, intratumor phenomena. The cancer itself was believed to be a tissue that was simple in its composition but extremely difficult to eradicate.

Therefore, model systems were developed for experimental therapeutics studies that employed murine tumors selected for their invariability, their constancy, and their homogeneity. It was believed that these homogeneous tumor models would provide relevant material for drug studies designed to identify agents active against human cancers. Thus many of us grew up in the world dominated by the L1210 and P388 mouse leukemias and similar murine neoplasms. These models provided investigators with reproducible results and experimental systems where attention was focused on drugs and not on the tumor.[15-17] The cancer cells were merely tools, targets, to aid us in our goal to establish chemotherapy protocols effective for the treatment of human cancers.

It is certainly true that L1210 and other murine tumors, usually leukemias and sarcomas, were quite instrumental in identifying drugs active against malignant neoplasms, particularly hematogenous neoplasms. It is also true that the extensive efforts made using these systems tended to mask the real nature of the tumor itself. The heterogeneity of most cancers remained hidden behind a screen of homogeneous L1210 cells. The workers using the murine leukemias in pharmacological, biochemical, and other studies far outnumbered the few investigators struggling to understand the biology of tumors. Consequently, the realization by a few that mouse tumors could be quite heterogeneous was opposed by a general belief that all cancers, like L1210, were homogeneous tissues.

III. HETEROGENEITY WITHIN MURINE TUMORS

Over the past 30 years, reports have continued to appear that indicated that diversity was indeed a characteristic of mouse neoplasms. Investigations of mouse mammary tumor biology led Leslie Foulds to postulate his concept of tumor progression. The collection of four articles in 1956 in one issue of the *Journal of the National Cancer Institute* serves as a classical treatise defining the existence and principles of intratumor heterogeneity.[18-21] Fould's perception of tumor progression necessarily implied and predicted heterogeneity within neoplasms. Progression as defined by Foulds means simply that there are sequential, independent and lasting, heritable changes occurring for many properties in cancers. Each change, each characteristic, would be vested in a subset of cells. Ultimately, variant clones that are the most aggressive, the most autonomous, and/or the most malignant would populate the tumor. The host must eventually succumb to this process of lethal progression. Kerbel[22] has stated that the importance of Fould's concept cannot be overestimated. This is absolutely correct. Tumor progression, the dynamic evolution of the neoplasm as evidenced by the appearance and selection of emerging subpopulations, is the heart and soul of cancer biology. It is also, more than any other factor, responsible for our continued inability to cure the majority of patients suffering from neoplastic disease.

IV. PHENOMENOLOGY OF MURINE TUMOR HETEROGENEITY

Many of the early reports on tumor heterogeneity described diversity in murine tumors rather than human tumors. Much of this work was done with tumor lines, either cultured continuously in the incubator, or passaged from mouse to mouse as transplantable neoplasms. Consequently, there was always the possibility that diversity within a mouse tumor line propagated in the laboratory over a period of years could arise as an artifact of continuous passage. However, other studies have been performed on spontaneously arising murine neoplasms, or newly induced mouse tumors. Results with these "recent" murine cancers clearly confirm findings obtained from established lines.[23] The relevance of these demonstrations of tumor heterogeneity in the mouse for the disease in man can be questioned. However, as will be seen in Chapters 2 and 5, tumor heterogeneity also occurs frequently in all the major solid human tumors as well. Thus, experimental work with murine tumors has provided an introduction and an example for the critical work performed primarily in the past 10 years with human cancers.

A. Immunological Heterogeneity
Prehn[24] demonstrated in 1970 that a methylcholanthrene-induced murine sarcoma in early passage contained two antigenically distinct subpopulations. Subsequent work by Pimm and Baldwin[25] supported Prehn's finding. When cell lines were established from

recurrent tumors at sites of excision of primary rat sarcomas, the primary and recurrent lines were shown to be immunologically distinct. Other workers have demonstrated that metastases of murine tumors may have antigens different from those found on the primary tumor.[26] AKR mouse thymic lymphomas contain a majority clone, and one or more minority clones that are immunologically different from the dominant subpopulation.[27] The examples cited above for the most part involved studies where transplantation type experiments and other immunological assays demonstrated the inability of one clone from a tumor to protect against the growth of another variant from the same original mouse neoplasm. However, another possibility is that a subpopulation will evolve during the progression of the tumor, which has lost its immunogenicity. Such a clone would have a survival advantage over other antigenic variants because it would presumably escape the host (specific) immune system.[28]

Heppner and her colleagues have also explored antigenic heterogeneity using subpopulations isolated from a spontaneously arising mouse mammary carcinoma.[29] These clones had significantly different levels of expression of mouse mammary tumor virus (MMTV) antigen.[29] Moreover, the expression of MMTV antigen could be differentially modulated by iododeoxyuridine.[30] Using classical techniques, Miller and Heppner[31] demonstrated that the subpopulations in this system could differ in their respective abilities to immunize as well as to serve as targets in an immune response.

The evidence is quite clear that murine tumors can generate variants with altered antigenic profiles. Such an altered immunogenicity confers a survival advantage on the variant clone in many cases. As will be seen, antigenic heterogeneity also exists with human tumors, and has important implications both for the detection and the treatment of these cancers.

B. Heterogeneity for Chromosomal and Biochemical Markers

Karyotypic diversity among cells within a neoplasm has provided incontestable evidence for intratumor heterogeneity. Work with mouse mammary tumor subpopulations has demonstrated large differences in model chromosome number and range for these clones.[29] Mitelman et al.[32] and others[33-34] have also described karyotypic heterogeneity for murine tumors. There are a number of reports documenting karyotypic diversity for human cancers; these are covered in the next chapter.

Chromosomal difference among subpopulations, particularly the acquisition of extra numbers of chromosomes by tumor cell variants, is perfectly consistent with the concept of progression. It is also consistent with the genetic basis for progression that has been proposed by Nowell,[35] as will be discussed in Chapter 3. It is entirely reasonable that a more anaplastic, aggressive, progressed tumor should also contain cells with an abnormal number of chromosomes, as well as alterations in the structure of the chromosomes themselves. Such changes would be compatible with, and are perhaps necessary for, the continued evolution of the cancer as an autonomous growth that increasingly threatens its host.

Heterogeneity within individual mouse neoplasms for a variety of biochemical markers has been reported by a number of laboratories. Variability between subpopulations for melanin production in hamster and mouse melanomas has been demonstrated by several groups.[36-38] Differential expression of tumor cell plasma membrane receptors for lectins has also been shown in different studies.[39-41] An investigation by Sluyser and Van Nie has revealed that hormone-induced mammary carcinomas in GR mice contain both estrogen-receptor positive and estrogen-receptor negative cells.[42] We shall see in the next chapter that there is compelling evidence for estrogen-receptor heterogeneity in human breast cancer.

It is perhaps evident at this point that murine tumors are often heterogeneous, and

INTRANEOPLASTIC DIVERSITY

homogeneous
neoplasm

neoplasm heterogeneous
for phenotype "A"

neoplasm heterogeneous
for phenotype "B"

neoplasm heterogeneous
for phenotype "C"

FIGURE 1. Intraneoplastic diversity must be defined for the phenotype in question. The heterogeneity pattern can be quite complex, with a shuffling of subpopulations depending on the characteristic being determined. Here, phenotype A could represent pigment production, with some melanoma cells containing more melanin than others. However, if the same tumor is examined for sensitivity to ionizing radiation, a different subset of cells is now more radioresistant (phenotype B) compared to the pigmented subset. Finally, another different subset of cells has evolved which displays a unique antigenic pattern (phenotype C), distinct from all other cells in the neoplasm. A cell displaying the "A" phenotype may or may not display the "B" or the "C" phenotype.

that they can show variability for a variety of phenotypic characteristics (Figure 1). In fact, one must be careful to specify the phenotype being evaluated when one speaks of a tumor as being heterogeneous.[43] Intraneoplastic diversity has been demonstrated for murine tumors for just about any phenotypic trait that can be assessed, from simple morphology to complex metabolic and antigenic profiles (Table 1). Heterogeneity within animal tumors for metastatic potential and variation between primary tumor and their metastases for response to treatment modalities are now compellingly documented and will be discussed in detail later. Such findings have assumed an even greater relevance now that we have discovered that the same variability for these diverse phenotypes exists within human cancers. Accordingly, work done with murine models of intratumor heterogeneity should help us better understand the phenomenon as it occurs in the human disease. However, research on human tumor diversity will also be necessary before we can come to a fuller appreciation of the implications of intraneoplastic diversity for cancer patients.

Table 1
VARIOUS PHENOTYPES FOR WHICH HETEROGENEITY HAS BEEN
DEMONSTRATED IN MURINE TUMOR LINES

Phenotypic trait	Experimental model
Culture morphology	Many tumor cell lines
Tissue histology	Notably mouse mammary tumors
Karyotype	Mouse mammary and other tumor cell lines
Growth rate	Various murine neoplasms (in vivo) as well as a number of culture cell lines
Tumor potential	Various tumor types including mouse melanoma and mammary carcinoma
Antigenicity	Many models, using transplantation studies, response to antigenic stimulation, sensitivity to antibody or immunocyte attack, and virus expression
Biochemical markers	Melanin content in melanomas, estrogen receptor levels in mammary tumors, and lectin receptors (in B16 and lymphosarcomas) are among the markers that have been assayed
Invasiveness	B16, using various barriers
Metastasis	Many models including the B16, the mouse mammary tumor system of Heppner and colleagues, and the lymphosarcoma system of Nicolson and co-workers; studies have used in vitro lines and in vivo tumors
Treatment response	Many models; modalities examined have included a large number of drugs, ionizing radiation, and response to differentiating agents

Note: A tumor or tumor cell line can be heterogeneous for many phenotypic characteristics. Therefore heterogeneity must be defined for the phenotype specified. Murine tumors (and their metastases) have been shown to be heterogeneous for many phenotypic traits; indeed, where differences have been sought, they have been found. A partial list, which could be lengthened considerably, is shown in the table.

C. Heterogeneity for Invasive and Metastatic Potential

This topic is so important that it has been covered in two separate chapters (Chapters 4 and 9). It is worth mentioning at this point, however, that the classical paper on tumor heterogeneity by Fidler and Kripke,[44] which stimulated the entire field, dealt with heterogeneity for metastatic potential among clones of the B16 murine melanoma. These investigators elegantly demonstrated that a parent B16 cell line could be cloned to provide subpopulations that have distinct differences in their ability to colonize lungs following i.v. injection into recipient mice. Some clones had a high lung colonizing capability whereas others had a markedly reduced ability to produce tumors in lung tissue following tail vein inoculation. Thus in one experiment, simple in design, the heterogeneity of a tumor for that most deadly characteristic, the ability to disseminate, was convincingly demonstrated. The clinical implications of that study were immediately apparent, and the report by Fidler and Kripke was correctly perceived to be of extreme importance to our understanding of the pathogenesis and treatment of neoplastic disease.

Heterogeneity for metastatic potential in human tumor lines has also been demonstrated. Studies on the A375 human melanoma line were conducted by Kozolowski et al., who cloned the parent line and injected the parental and clonal populations tail vein into 3- to 4-week-old nude mice. Clonal heterogeneity was demonstrated for lung colonization by the melanoma cells. When the pulmonary metastases were removed and cell lines established from them, the cells from the secondary tumors were more metastatic than parent A375 cells when injected i.v. or s.c. into young nude mice.[45] Kerbel et al. used an innovative in vitro/in vivo selection sequence to isolate variant subpopulations from the human melanoma MeWo line. The variant cells had a significantly greater propensity for lung colonization in nude mice following i.v. inoculation than did the parental cells.[46] Thus the same manipulations that have been frequently and successfully carried out to select murine tumor cells with metastatic potential have been applied to human melanoma cells as well.

D. Heterogeneity for Response to Treatment Modalities

Variability among subpopulations for the "treatment response phenotype" is really the bottom line for the clinical significance of tumor heterogeneity. Without this type of diversity, all the other differences would be of only biological interest. If all clones including those that had metastasized "saw" drugs, radiation, heat, immunotherapeutic agents, and hormones identically, and responded to them in an identical manner, the other differences would not be important to the clinician. He could treat all heterogeneous tumors, all subpopulations, in the same way, regardless of their other variations. And we would be back to the L1210 philosophy again: find the right drug(s) for the (homogeneous) tumor. However, as might be expected, there is an incredible heterogeneity for response to various treatment modalities among clones from both murine and human tumors. Chapters 8 and 9 will be devoted to this topic.

V. SUMMARY

Murine neoplasms have given us a window into the biology of cancer. A dynamic world of tumor progression has been revealed which has gradually led us to an appreciation of the complexity of murine and human malignancies. We have learned from Foulds and from others including Fidler and Kripke that tumors are evolving in a dynamic manner to progressed forms able to metastasize and spread to distant organs. This progression is due to the presence of multiple clones within single tumors. Some of these clones have the ability to disseminate, and all of these subpopulations respond differently and unpredictably to therapeutic modalities. The clinical implications of these findings are ominous, and will be explored more fully throughout this volume.

REFERENCES

1. Dunn, T., Morphology of mammary tumors in mice, in *Physiopathology of Cancer,* 2nd ed., Homburger, F. and Fishman, N. H., Eds., Paul B. Hoeber, New York, 1959, 38.
2. Henson, D. E., Heterogeneity in tumors, *Arch. Pathol. Lab. Med.,* 106, 597, 1982.
3. Virchow, R., *Die Krankhaften Geschwulste,* Vol. 2, August Hirschwald, Berlin, 1864.
4. Bigner, D. D., Pedersen, H. B., Bigner, S. H., and McComb, R., A proposed basis for the therapeutic resistance of gliomas, *Semin. Neurol.,* 1, 169, 1981.
5. Stroebe, H., Über entestehung und bau der gehirnglioma, *Beitr. Pathol. Anat. Alla. Pathol.,* 18, 405, 1895.
6. Penfield, W., Classification of brain tumors and its practical application, *Br. Med. J.,* 1, 337, 1931.
7. Bailey, P. and Cushing, H., Eds., *A Classification of Tumors of the Glioma Group,* Lippincott, Philadelphia, 1926.
8. del Rio-Hortega, P., Ed., *The Microscopic Anatomy of Tumors of the Central and Peripheral Nervous System,* Charles C Thomas, Springfield, 1962.
9. Courville, C. R., Cell types in the gliomas, *Arch. Path.,* 10, 649, 1930.
10. Foulds, L., The experimental study of tumor progression: a review, *Cancer Res.,* 14, 337, 1954.
11. Calabresi, P., Dexter, D. L., and Heppner, G. H., Clinical and pharmacological implications of cancer cell differentiation and heterogeneity, *Biochem. Pharmacol.,* 28, 1933, 1979.
12. Heppner, G. and Miller, B. E., Biological variability of mouse mammary neoplasma, in *Design of Models for Testing Cancer Therapeutic Agents,* 3rd ed., Fidler, I. J. and White, R. J., Eds., Van Nostrand Reinhold Co., New York, 1982, chap. 4.
13. Ewing, J., *Neoplastic Diseases,* 3rd ed., W. B. Saunders, Philadelphia, 1928, chap. 4.
14. Paget, S., The distribution of secondary growth in cancer of the breast, *Lancet,* 1, 571, 1889.
15. Heidelberger, C. and Ansfield, F. J., Experimental and clinical use of fluorinated pyrimidines in cancer chemotherapy, *Cancer Res.,* 23, 1226, 1963.
16. Venditti, J. M., The model's dilemma, in *Design of Models for Testing Cancer Therapeutic Agents,* 3rd ed., Fidler, I. J. and White, R. J., Eds., Van Nostrand Reinhold Co., New York, 1982, chap. 7.

17. Kline, I., Woodman, R. J., Gang, M., Cysyk, R. L., and Venditti, J. M., Value of cytosine arabinoside (NSC-63878) plus 5-(3,3-dimethyl-1-triazeno) imidazole-4-carboxamide (NSC-45388) therapy in advanced murine L1210 leukemia and enhancement of the combination with sequential doses of methotrexate (NSC-740) or 1,3-bis (2-chloroethyl)-1-nitrosourea (BCNU; NSC-409962), *Cancer Chemother. Rep.*, 57, 291, 1973.
18. Foulds, L., The histologic analysis of mammary tumors of mice. I. Scope of investigations and general principles of analysis, *J. Natl. Cancer Inst.*, 17, 701, 1956.
19. Foulds, L., The histologic analysis of mammary tumors of mice. II. The histology of responsiveness and progression. The origin of tumors, *J. Natl. Cancer Inst.*, 17, 713, 1956.
20. Foulds, L., The histologic analysis of mammary tumors of mice. III. Organoid tumors, *J. Natl. Cancer Inst.*, 17, 755, 1956.
21. Foulds, L., The histologic analysis of mammary tumors of mice. IV. Secretion, *J. Natl. Cancer Inst.*, 17, 783, 1956.
22. Kerbel, R. S., Implications of immunological heterogeneity of tumors, *Nature (London)*, 280, 358, 1979.
23. Kripke, M. L., Gruys, E., and Fidler, I. J., Metastatic heterogeneity of cells from an ultraviolet light-induced murine fibrosarcoma of recent origin, *Cancer Res.*, 38, 1962, 1978.
24. Prehn, R. T., Analysis of antigenic heterogeneity within individual 3-methylcolanthrene-induced mouse sarcomas, *J. Natl. Cancer Inst.*, 45, 1039, 1970.
25. Pimm, M. V. and Baldwin, R. W., Antigenic differences between primary methylcholanthrene-induced rat sarcomas and post-surgical recurrences, *Int. J. Cancer*, 20, 37, 1977.
26. Sugarbaker, E. V. and Cohen, A. M., Altered antigenicity in spontaneous pulmonary metastases from an antigenic murine sarcoma, *Surgery*, 72, 155, 1972.
27. Olsson, L. and Ebbesen, P., Natural polyclonality of spontaneous AKR leukemia and its consequences for so-called specific immunotherapy, *J. Natl. Cancer Inst.*, 62, 623, 1979.
28. Klein, G. and Klein, E., Immune surveillance against virus-induced tumors and nonrejectability of spontaneous tumors: contrasting consequences of host versus tumor evolution, *Proc. Natl. Acad. Sci. U.S.A.*, 74, 2121, 1977.
29. Dexter, D. L., Kowalski, H. L., Blazar, B. H., Fligiel, Z., Vogel, R. and Heppner, G. H., Heterogeneity of tumor cells from a single mouse mammary tumor, *Cancer Res.*, 38, 3174, 1978.
30. Hager, J. C. and Heppner, G. H., Heterogeneity of expression and induction of mouse mammary tumor virus antigens in mouse mammary tumors, *Cancer Res.*, 42, 4325, 1982.
31. Miller, F. R. and Heppner, G. H., Immunologic heterogeneity of tumor cell subpopulations from a single mouse mammary tumor, *J. Natl. Cancer Inst.*, 63, 1457, 1979.
32. Mitelman, F., Mark, J., Leran, G. and Levan, S., Tumor etiology and chromosome pattern, *Science*, 176, 1340, 1972.
33. Makino, S., The chromosome cytology of the ascites tumor of rats, with special reference to the concept of the stemline cell, *Intern. Rev. Cytol.*, 6, 26, 1957.
34. Suzuki, N., Frapart, M., Fudina, D. J., Meistrich, M. L., and Withers, H. R., Cell cycle dependency of metastatic lung colony formation, *Cancer Res.*, 37, 3690, 1977.
35. Nowell, P. C., The clonal evolution of tumor cell populations, *Science*, 194, 23, 1976.
36. Gray, J. M. and Pierce, G. B., Relationship between growth rate and differentiation of melanoma *in vivo*, *J. Natl. Cancer Inst.*, 32, 1201, 1964.
37. Fidler, I. J. and Hart, I. R., Biological and experimental consequences of the zonal composition of solid tumors, *Cancer, Res.*, 41, 3266, 1981.
38. Dexter, D. L., Lee, E. S., DeFusco, D. J., Libbey, N. P., Spremulli, E. N., and Calabresi, P., Selection of metastatic variants from heterogeneous tumor cell lines using the chicken chorioallantoic membrane and nude mouse, *Cancer Res.*, 43, 1733, 1983.
39. Raz, A., McLellan, W. L., Hart, I. R., Bucana, C. D., Hoyer, L. C., Sela, B. A., Dragsten, P., and Fidler, I. J., Cell surface properties of B16 melanoma variants with differing metastatic potential, *Cancer Res.*, 40, 1645, 1980.
40. Reading, C. L., Belloni, P. N., and Nicolson, G. L., Selection and in vivo properties of lectin-attachment variants of malignant murine lymphosarcoma cell lines, *J. Natl. Cancer Inst.*, 64, 1241, 1980.
41. Brunson, K. W. and Nicolson, G. L., Selection and biologic properties of malignant variants of a murine lymphosarcoma, *J. Natl. Cancer Inst.*, 61, 1499, 1978.
42. Sluyser, M. and Van Nie, R., Estrogen receptor content and hormone-responsive growth of mouse mammary tumors, *Cancer Res.*, 34, 3253, 1974.
43. Dexter, D. L. and Calabresi, P., Intraneoplastic diversity, *Biochim. Biophys. Acta*, 695, 97, 1983.
44. Fidler, I. J. and Kripke, M. L., Metastasis results from pre-existing variant cells within a malignant tumor, *Science*, 197, 893, 1977.

45. Kozolowski, J. M., Hart, I. R., Fidler, I. J., and Hanna, N., A human melanoma line heterogeneous with respect to metastatic capacity in athymic nude mice, *J. Natl. Cancer Inst.*, 72, 913, 1984.
46. Kerbel, R. S., Man, M. S., and Dexter, D., A model of human cancer metastasis: extensive spontaneous and artificial metastasis of a human pigmented melanoma and derived variant sublines in nude mice, *J. Natl. Cancer Inst.*, 72, 93, 1984.

breast cancer are now being interpreted as conclusive evidence that human breast cancers demonstrate intratumor and interlesional heterogeneity.

As might be expected from analyses on human breast carcinomas such as described above, cell lines established from breast cancers from patients have also proven to be heterogeneous. One demonstration of variability for a property often equated with tumor progression, anchorage independence, has recently been reported. Hancock and Smith have presented good evidence that heterogeneity for the anchorage-independence phenotype is generated rapidly in cultured Hs578T human mammary cancer cells.[9] Such instability for an aggressive phenotype in breast tumors in women could lead to the evolution of more anaplastic, and perhaps more metastatic clones, in these patients.

B. Lung Cancer

Carcinoma of the lung has emerged as the major problem among all types of cancer. Fully 25% of all cancer-related deaths in the U.S. are due to lung cancer, and this proportion of lethalities will increase as the mortality rate among women with bronchogenic carcinoma continues to rise.[1] This will happen because of the increased cigarette smoking among women since the second world war. Indeed, clinicians and health care officials are quite justified in stating that we are facing an epidemic of lung cancer in this country.

Lung cancer, in general, has been refractory to treatment regardless of the form in which it occurs. Small cell carcinoma of the lung is considered to be disseminated at the time it is diagnosed, and chemotherapy, often with radiotherapy, is the treatment of choice.[10,11] Small cell lung cancers (SCLC) are quite radiosensitive, but they are not usually radiocurable. The nonsmall cell lung carcinomas (NSCLC) include large cell lung cancer, and squamous cell carcinoma and adenocarcinoma of the lung. These neoplasms are currently not being successfully treated in many patients.

One would be tempted to speculate that perhaps human lung tumors are often each heterogeneous neoplasms, and the difficulty clinicians are encountering in eradicating these cancers may be due to intraneoplastic diversity in lung carcinomas. There are good indications that this is the case. Indeed, some of the best evidence obtained to date showing that individual human solid tumors are heterogeneous has been provided by studies of lung cancer.[12-14]

Olsson et al. have established cloned cell lines and subclones of these lines from fresh biopsies of human small cell and squamous cell lung cancers.[15] Phenotypic diversity among subclones from a given line was demonstrated for karyotype, growth rate, morphology, and clonogenicity in soft agar. Importantly, intratumoral antigenic heterogeneity was shown for several distinct monoclonal antibodies for human lung tumor cells. The antibody binding patterns were different for different subclones from the same tumor, and for clonogenic vs. nonclonogenic cells from a given carcinoma. In some cases, clonogenic cells bound antibody, whereas in other instances they did not.[15] The implications of these results for tumor diagnosis are obvious.

The use of flow cytometry has provided a window through which one can look at intratumor heterogeneity as it exits in the patient's neoplasm.[16-19] This technique provides compelling data regarding karyotypic and kinetic heterogeneity among cell subpopulations in single tumors and is discussed in detail in Chapter 5. The method utilizes samples directly taken from surgical or biopsy materials and avoids any artifacts that might arise in the propagation of human tumor cells in culture or in nude mice. Flow cytometry has been applied to both small cell and nonsmall cell carcinomas of the lung and has provided strong evidence for the existence of heterogeneity within individual lung cancers.

A group from Italy has reported that 10 of 11 human nonsmall cell lung carcinomas were karyotypically heterogeneous as shown by flow cytometric data.[20] These ten cancers contained at least one aneuploid subpopulation in addition to a diploid population of cancer cells. Furthermore, the proportions of aneuploid and diploid cells were not constant within a single heterogeneous lung neoplasm, but changed depending on what region of the tumor was sampled.

Danish investigators have examined SCLC for the presence of multiple clones within single lesions using flow cytometry. These workers looked at metastases from a series of 29 patients with small cell lung cancer.[21] In six patients, karyotypic heterogeneity was demonstrated either within a single metastasis or between two metastases from the same person. In these six cases, the flow cytometric profiles showed that tumor samples contained both diploid and aneuploid cancer cells. Moreover, it was concluded that this percentage of heterogeneous tumors (21%) is most likely an underestimate because the method cannot resolve aneuploid subpopulations comprising less than about 10% of the total tumor mass.

A fascinating story is now emerging from studies looking at other phenotypic characteristics of lung cancer as markers for heterogeneity. The parameters studied are essentially differentiation markers including histological and enzymatic properties that enable a lung carcinoma to be classified as belonging to a particular type, i.e., adenocarcinoma, etc. Most of this work has centered around clinical and experimental analyses of small cell lung cancer.

The clinical data derive from studies of primary and disseminated SCLC in patients at the time of presentation and/or at autopsy. Combined subtypes of SCLC and NSCLC (squamous carcinoma and adenocarcinoma), according to the World Health Organization (WHO) classification system, appear in less than 1% of patients presenting with SCLC. However, the combined type is observed in over 33% of treated cases at autopsy.[22,23] One study documented that 35 of 97 SCLC cases at autopsy showed the presence of NSCLC types. Fifteen of these contained large cells and twenty had adeno, squamous, and large cell components.[22] In another investigation, analyses were done on 40 autopsies performed on patients with an original diagnosis of SCLC. There were six instances of mixed SCLC and large cell cancer, and five cases where only NSCLC histologies could be obtained. Biochemical profiles obtained on these lesions, using histaminase and L-dopa decarboxylase as markers, supported the histological evidence.[23] The presence of large cell plus small cell types (WPL intermediate 22/40 classification) in a tumor confers poorer prognosis on the patient compared to histologically homogeneous SCLC.[12]

Histological heterogeneity in lung cancer has also been studied using another intermediate classification where WPL-21 is the SCLC type, WPL-22 is the NSCLC type, and WPL-21/22 is a mixed tumor. The mixture subtype occurs in about 7% of patients diagnosed with SCLC. Furthermore, when multiple biopsies from each of a series of WPL-21/22 individuals were examined, there was a discordance of subtype assignment between primary and secondary lesions from a patient in 26% of cases.[24]

Laboratory studies have been performed with cell lines established from SCLC specimens. One such line (NCI-H128) was cloned in soft agar, and individual colonies were assayed for levels of the peptide hormones ACTH, PVP, and calcitonin. The parent cell line had significant levels of all three markers. There was clonal heterogeneity of expression of the hormones, with individual clones expressing one, two, or all three polypeptides.[12] When several SCLC lines were studied under conditions of continuous growth in culture or as xenograft tumors in nude mice, the appearance of cells with transitional and large cell morphology has been documented. They are related to SCLC by virtue of the presence of a marker chromosome [del (3) (p14-23)] specific for SCLC, high levels of BB creatinine kinase, and moderate levels of a neuron-specific enolase.

However, they differ from SCLC in that these NSCLC type cells have shorter doubling times, clone better in agar, and do not express L-dopa decarboxylase.[12,25,26]

There are two "favored" hypotheses that have been developed to explain the heterogeneity observed in SCLC in patient material and in derivative cell lines.[12,14] The first suggests that treatment has selected for a minority, resistant subtype that becomes more important, even predominant, by the time the patient succumbs to his disease. Most of the autopsy tissue examined to date has come from treated patients.

The second hypothesis proposes that the SCLC cells have the ability to differentiate to NSCLC forms. Thus the small cell carcinomas contain stem cells (or are wholly stem cell tumors) that can produce the other varieties. Two groups have used electron microscopy to demonstrate the presence of small and squamous cell components in untreated patients with a SCLC diagnosis according to light microscopic criteria.[27,28] These data favor a conversion rather than a selection event. More definitive data recently have been reported from Goodwin et al.[29] These workers have established and cloned cell cultures of human large cell undifferentiated lung cancer from a patient with mixed SCLC and NSCLC. The cloned lines expressed cell surface proteins that were characteristic of SCLC and NSCLC. They interpreted their findings to suggest that "small cell lung carcinoma and a form of large cell carcinoma appear linked through a continuum of differentiation events." Since the prognosis associated with the appearance of NSCLC types is poorer than with "homogeneous" SCLC, the differentiation process is not directed towards a more benign, mature end cell, and is therefore not differentiation in the sense that developmental biologists use the term. Rather, it would seem to be a case of progression, as was discussed in Chapter 1, and perhaps should be more appropriately called a dedifferentiation process characterized by a loss of neuroendocrine differentiated features.[12,29]

The hypothesis that SCLC can progress to a more poorly differentiated type with a less favorable clinical prognosis is supported by a study by Goodwin and Baylin.[30] These workers demonstrated that the human SCLC line OH-1 in a 16-month period lost its ultrastructural and biochemical markers of neuroendocrine differentiation and acquired a concomitant 1- to 2-log increase in resistance to ionizing radiation. The authors noted the parallelism with the clinical behavior of SCLC.

Much work remains to be done in dissecting the phenomenon of heterogeneity in lung cancer. The origins of this diversity remain to be elucidated. In particular, the evolution occuring in SCLC is currently not understood, and provides a fascinating challenge to medical and experimental oncologists.[31] One important finding that has emerged from the work with human lung cancer heterogeneity is that it has an important bearing on the fate of the patient. There is a definite association between the extent and pattern of heterogeneity and the patient's prognosis.

C. Colorectal Cancer

Carcinomas of the colon and rectum have only been surpassed recently by lung cancers as the leading cause of death among neoplasms in the U.S.[1] This tumor type has been resistant to the efforts of medical oncologists and radiotherapists. If surgery is not successful in completely eradicating this disease, the patient frequently dies of disseminated cancer; the liver and the lungs are favorite organs for colonization.[32]

As is the case in breast and lung cancers, there is good evidence for heterogeneity within individual colon tumors at the time of their excision. Furthermore, the existence of subpopulations of neoplastic cells within these tumors again seems to be the usual state of affairs rather than the exception. Thus, there is indication that the intraneoplastic diversity occurring within colorectal cancers is a primary factor in our general inability to effectively treat tumors of this class.

Peterson et al.[33] have studied six consecutive cases of colorectal carcinomas for karyotypic heterogeneity, immediately upon surgical removal. Five of these tumors contained hyperploid subpopulations as well as a majority diploid subpopulation. In another investigation, a series of carcinomas weere studied; two colon cancers were among this series of tumors. Heterogeneity in the response of portions of each tumor to a battery of chemotherapeutic drugs was demonstrated.[34]

In our own laboratory we have studied in detail two heterogeneous colon cancer systems. The findings will be reported elsewhere in this volume (Chapter 8), but a brief description can be provided here. The heterogeneous DLD-1 colon carcinoma was reported to be histologically heterogeneous in the original pathology report on the patient. The primary tumor removed at surgery consisted of anaplastic areas interspersed with moderately differentiated regions containing acinar structures.[35,36] Subsequent studies confirmed that the DLD-1 tumor is heterogeneous for many phenotypic characteristics. Two metastases were obtained from another patient with primary colon cancer. One was a secondary tumor growing in the omentum and the other lesion was a tumor deposit in the ovary. Analysis done on xenografted tumors established in nude mice from biopsy material from these metastases showed that they differentially expressed several marker antigens. Analysis of cell lines established from these secondary tumors revealed still other differences.[8,37,38]

Antigenic heterogeneity has also been reported for the LoVo human colon cancer cell line using monoclonal antibodies. Twelve subclones of the LoVo parent line were tested for reactivity with a panel of monoclonal antibodies, and variable expression of antigens was demonstrated.[39]

An interesting report has appeared from Sweden; a group of investigators studied 21 endocrine tumors of the rectum that produced various pancreatic and gut neurohormonal polypeptides.[40] Immunocytochemical analyses allowed subsets of cells within individual endocrine rectal carcinomas to be identified according to whether they contained specific marker polypeptides. Eleven of the twenty-one tumors contained more than one specific peptide-expressing cell subpopulation. Furthermore, the authors postulated that the use of other antisera specific for different peptides would probably reveal the presence of still other subpopulations in those ten tumors that were homogeneous in their study.

D. Neuroblastoma

The extensive variability that has been documented within murine malignant testicular neoplasms is due to differentiation processes taking place within teratomas. There is a well-known parallel in human neoplastic disease — pediatric neuroblastoma. The clinical experience with the differentiation of this tumor type goes back more than half a century. Stem cells in neuroblastomas can differentiate to produce an intermediate type of malignancy called a ganglioneuroblastoma, and can go on to differentiate to produce a benign ganglioneuroma.[41-44] Thus, tumors of this type can contain malignant and differentiated cells (derived from the neoplastic cells), and are certainly heterogeneous cancers. In cases where the tumor converts to a totally benign form, the patient is cured of his disease.

E. Medullary Thyroid Carcinoma

Another tumor type, medullary thyroid carcinoma (MTC), has proven to be an excellent model for the study of human intraneoplastic diversity. The work on MTC heterogeneity has been primarily done by Baylin and co-workers.[45,46] Baylin and Mendelsohn have reviewed features of MTC that make this cancer quite suitable for heterogeneity studies.[46] First, hereditary forms of MTC exist.[47] Second, the normal parent cell for MTC has been identified.[48] Third, a biochemical phenotype for the MTC parent cell has been worked out in several laboratories.[49-51] Fourth, there is a large range of clinical behavior associated with MTC.[14,52-54]

The data obtained with specimens of MTC obtained from patients at various stages of each individual's disease, including disseminated MTC, reveal a striking pattern of biochemical heterogeneity. Calcitonin expression decreases, and a heterogeneous pattern of intratumor calcitonin expression becomes more evident with progressive disease. There is a relative increase in expression of L-dopa decarboxylase and diamine oxidase concomitant with disease progression and reduced levels of calcitonin. The worsening clinical picture (prognosis) parallels the biochemical profile.[46,53-57]

The evolution of MTC thus in some ways can be compared to that occuring in SCLC tumors that produce NSCLC components. A primitive stem cell in MTC undergoes a series of controlled, or programmed changes to convert to other cell types, producing finally a very progressed tumor with reversal of the marker ratio profile. As in SCLC, these events can be viewed as a dedifferentiation process whereby a more anaplastic, poorly differentiated tumor develops with time. One has an anomaly here, because we are now speaking of the dedifferentiation of a stem cell type. The supposedly primitive cell, of which the original (biochemically homogeneous?) neoplasm is comprised, "differentiates" according to a fixed program into less mature forms. This is diametrically opposed to what is observed in teratoma or neuroblastoma. It is also quite different from the random changes leading to progressed heterogeneous cancers that were predicted by Foulds[58] and defined by Nowell[59] (see Chapter 3).

Certainly much important information on the origin, maintenance, and implications of human tumor heterogeneity will come from further studies with MTC. This tumor type remains one of the best models for studying human intraneoplastic diversity.

F. Other Tumors

The five tumor types discussed separately were included because of their extreme clinical relevance (breast, lung, and colorectal carcinoma) or because of their importance in illustrating important aspects of the phenomenon of intratumor heterogeneity (neuroblastoma and medullary thyroid carcinoma). Information on intraneoplastic diversity for these cancers has been provided by direct studies on tissues or fluids obtained from patients at surgery, biopsy, or autopsy.

One can conclude that the case for human tumor heterogeneity has been made convincingly during the past 5 to 10 years. The examples cited earlier in this chapter support the belief of most medical and experimental oncologists that intraneoplastic diversity is a major problem in treating patients with solid tumors. Moreover, it may be worth pointing out that an impressive documentation of intratumor diversity exists for many others types of human cancers as well. Again, much of the evidence derives from studies of patient material conducted quickly after it has been collected. However, important data have also been obtained from cell lines and xenograft tumors propagated continuously in culture or in nude mice following establishment of a line in the laboratory from patient tissue. These findings are reviewed here so that researchers and clinicians not completely familiar with intratumor heterogeneity will appreciate that the phenomenon occurs across the spectrum of human cancers.

1. Ovarian Tumors

Advances in flow cytometric technology have permitted multiparameter analysis of cell preparations.[60,61] DNA content, several enzymes, and phagocytic activity can be measured simultaneously on a cell suspension using a cytofluorograph equipped with three photomultiplier tubes.[62] When human ovarian cancers were examined using this methodology, both biochemical and karyotypic heterogeneity could be demonstrated in the samples analyzed. The ascites fluid from one patient with ovarian cancer contained only diploid tumor cells, yet two subpopulations were identified based on alkaline phosphatase levels. Cells from an omental metastasis from the same patient were divided among three subset compartments based on alkaline phosphatase expression. Again, the ploidy values were normal for cells from the solid tumor deposit. However, a different picture emerged when an ovarian tumor, its omental implant, and ascites tumor cells, all from another patient, were assayed. The alkaline phosphatase profile was identical for cells from all three sources and indicated a homogeneous expression of this enzyme. However, two karyotypically distinct subpopulations of tumor cells were found in the ascites cells.[62] Such multiparameter analyses support the view that the pattern of heterogeneity will be distinct for each human tumor. Furthermore, characteristics (markers, phenotypes) will segregate independently within individual tumors and their metastases, confirming Foulds' findings with mouse mammary tumors.

Karyotypic heterogeneity was demonstrated in each of a series of 12 patients with ovarian cancer. Moreover, some of these tumors changed in the percentage of cells in each ploidy compartment following treatment.[34] Endometrial cancers from a series of 20 patients were studied to determine whether secretory conversion had taken place following hormonal therapy, and what percentage of cells in each tumor was affected. There was intratumor heterogeneity for secretory conversion in all the cases studied.[34]

2. Brain Tumors

Morphological and karyotypic heterogeneity has been demonstrated in human brain tumors. Eight cell lines established from individuals with gliomas each contained distinct subsets of cells according to chromosomal and morphological criteria.[63] Furthermore, subpopulations from these lines showed heterogeneous responses to several antineoplastic drugs.[64]

Workers studying the Duke-maintained culture D-54 MG of the S-172 human glioma cell line have cloned the parent line and isolated 8 distinct subpopulations.[65] These subpopulations and the parent glioma line were shown to be biochemically, morphologically, karyotypically, and antigenically distinct. Furthermore, there is no apparent correlation between the expression of any of these characteristics. A complex pattern of antigenic heterogeneity among the parent and cloned lines was demonstrated using antisera raised in three species against normal and tumor cells of the central nervous system. Differential sensitivity of clones to antisera against human fetal brain and other human glioma cell lines argues for a variable antigenic profile consistent with determinants heterogeneously present on glioma subpopulations. While heterogeneity in this system could be due to possible artifacts of tissue culture, the investigators concluded that the diversity was probably due to heterogeneity present in the original tumor. The results of this study illustrate quite nicely for a human tumor that heterogeneity must be defined for a particular phenotype; a shuffling of clones, depending on the phenotype studied, can occur in single heterogeneous glioma.

3. Other Examples

Testicular cancers in man have been known for decades to produce varying patterns of histology, including distinct patterns for different metastases from the same primary. Histopathologies documenting mixed elements of choriocarcinoma, seminoma, and embryonal carcinoma in the same individual are not uncommon. Indeed, testicular cancers have historically been cited by pathologists as a prime example of the variability that can exist within human tumors and among their metastases. Bronson et al.[66] have identified four histological types of neoplastic cells in a cell line established from a metastasis of a human testicular tumor.

Heterogeneity in human melanoma has also been reported. One of the earliest findings of intratumor heterogeneity involved four sublines established from a human melanoma. The four subpopulations were differentially sensitive to chemotherapeutic drugs.[67,68] Clones from the K 235 human melanoma cell line have also shown differential chemosensitivities to antineoplastic drugs.[69]

Heterogeneity for drug sensitivity has been demonstrated in non-Hodgkin's lymphomas. Lymph nodes removed from seven patients were divided into portions, and each portion was tested for sensitivity to each of four drugs. Heterogeneity in responses among portions from three of the seven patients was demonstrated.[70]

Several distinct neoplastic cell types have been defined when an alveolar rhabdomyosarcoma was examined ultrastructurally.[71] Karyotypic heterogeneity has been discovered in a careful study of the lineage of a human retinoblastoma cell line. The line established from the human tumor showed karyotypic diversity, and in succeeding years two sublines had developed. These produced tumors in nude mice with significantly different latency periods.[72] Examination by electron microscopy has revealed two types of cells in a primary clear cell thyroid carcinoma.[73]

III. SUMMARY

The evidence amassing from direct analysis of patient tumor material provides compelling support for the view that human tumors are frequently each heterogeneous tissues. Thus the dynamic evolution seen within experimental murine tumors also occurs within human cancers. The statement that most if not all human tumors will be heterogeneous for some phenotypic trait, and probably for many such characteristics, can now be made with confidence. The ramifications of these findings for the management of cancer patients are multiple and serious.

REFERENCES

1. American Cancer Society: 1984 Cancer Facts and Figures, A.C.S., New York.
2. Brennan, M. J., Donegan, W. L., and Appleby, D. E., The variability of estrogen receptors in metastatic breast cancer, *Am. J. Surg.,* 137, 260, 1979.
3. McGuire, W. L., Carbone, P. P., Sears, M. E., and Escher, G. C., Estrogen receptors in human breast cancer: an overview, in *Estrogen Receptors in Human Breast Cancer,* McQuire, W. L., Carbone, P. P., and Voller, E. P., Eds., Raven Press, New York, 1975, 1.
4. Lee, S. H., Cancer cell estrogen receptor of human mammary carcinoma, *Cancer,* 44, 1, 1979.
5. Poulsen, H. S., Jensen, J., and Hermansen, C., Human breast cancer: heterogeneity of estrogen binding sites, *Cancer,* 48, 1791, 1981.
6. Rosen, P. P., Menendez-Botet, C. J., Urban, J. D., Fracchia, A., and Schwartz, M. K., Estrogen receptor protein (ERP) in multiple tumor specimens from individual patients with breast cancer, *Cancer,* 39, 2194, 1977.
7. Bichel, P., Poulsen, H. S., and Anderson, J., Estrogen receptor content and ploidy of human mammary carcinoma, *Cancer,* 50, 1771, 1982.
8. Spremulli, E. N. and Dexter, D. L., Human tumor heterogeneity and metastasis, *J. Clin. Oncol.,* 1, 496, 1983.
9. Hancock, M. C. and Smith, H. S., Phenotypic variability in anchorage-independent growth by a human breast tumor cell line, *J. Natl. Cancer Inst.,* 72, 833, 1984.
10. Bunn, P. A., Jr. and Ihde, D. C., Small cell bronchogenic carcinoma: a review of therapeutic results, in *Lung Cancer,* Vol. 1, Livingston, R. B., Ed., Martinus Nijhoff, The Hague, 1981, 169.
11. Minna, J. D., Higgins, G. A., and Glatstein, E. J., Cancer of the lung, in *Cancer: Principles and Practice of Oncology,* DeVita, V. T., Hellman, S., and Rosenberg, S. A., Eds., Lippincott, Philadelphia, 1981, 396.
12. Minna, J. D., Carney, D. N., Alvarez, R., Bunn, P. A., Jr., Cuttitta, F., Ihde, D. C., Matthews, M. J., Oie, H., Resen, S., Whang-Peng, J., and Gazdar, A. F., Heterogeneity and homogeneity of human small call lung cancer, in *Tumor Cell Heterogeneity: Origins and Implications,* Vol. 4, 2nd ed., Owens, A. H., Coffey, D. S., and Baylin, S. B., Eds., Academic Press, New York, 1982, 29.

13. Baylin, S. B., Weisburger, W. R., Eggleston, J. C., Mendelsohn, G., Beaven, M. A., Abeloff, M. D., and Ettinger, D. S., Variable content of histaminase, L-dopa decarboxylase and calcitonin in small-cell carcinoma of the lung. Biologic and clinical implications, *N. Eng. J. Med.,* 299, 105, 1978.
14. Baylin, S. B. and Mendelsohn, G., Ectopic (inappropriate) hormone production by tumors: mechanisms involved and the biological and clinical implications, *Endocrinol. Rev.,* 1, 45, 1980.
15. Olsson, L., Sorensen, H. R., and Behnke, O., Intratumoral phenotypic diversity of cloned human lung tumor cell lines and consequences for analyses with monoclonal antibodies, *Cancer,* 54, 1757, 1984.
16. Beck, H. P., Evaluation of flow cytometric data of human tumors, *Cell Tissue Kinet.,* 13, 173, 1980.
17. Crissman, H. A., Kissane, R. J., Oka, M. S., Tobey, R. A., and Steinkamp, J. A., Flow microfluorometric approaches to cell kinetics, in *Growth Kinetics and Biochemical Regulation of Normal and Malignant Cells,* Drewinko, B. and Humphrey, R. M., Eds., William & Wilkins, Baltimore, 1977, 143.
18. Dean, P. N., A simplified method of DNA distribution analysis, *Cell Tissue Kinet.,* 13, 299, 1980.
19. Laerum, O. D. and Farsund, T., Clinical application of flow cytometry: a review, *Cytometry,* 2, 1, 1981.
20. Nervi, D., Gianna, B., Maisto, A., Mauro, F., Tirindelli-Danesi, D., and Starace, G., Cytometric evidence of cytogenetic and proliferative heterogeneity of human solid tumors, *Cytometry,* 2, 303, 1982.
21. Vindelov, L. L., Hansen, H. H., Christensen, I. J., Spang-Thomsen, M., Hirsch, F. R., Hansen, M., and Nissen, N. I., Clonal heterogeneity of small cell anaplastic carcinoma of the lung demonstrated by flow cytometric DNA analysis, *Cancer Res.,* 40, 4295, 1980.
22. Matthews, M. J., Effects of therapy on the morphology and behavior of small cell carcinoma of the lung — a clinicopathologic study, in *Lung Cancer: Progress in Therapeutic Research,* Muggia, F. and Rozencweig, M., Eds., Raven Press, New York, 1979, 155.
23. Abeloff, M. D., Eggleston, J. C., Mendelsohn, G., Ettinger, D., and Baylin, S. B., Changes in morphological and bischemical characteristics of small cell carcinoma of the lung, *Am. J. Med.,* 66, 757, 1979.
24. Carney, D. N., Matthews, M. J., Ihde, D. C., Bunn, P. A., Jr., Cohen, M. H., Makuch, R. W., Gazdar, A. F., and Minna, J. D., Influence of histologic subtype of small cell carcinoma of the lung on clinical presentation, response to therapy and survival, *J. Natl. Cancer Inst.,* 65, 1225, 1980.
25. Gazdar, A. F., Carney, D. N., and Minna, J. D., In vitro study of the biology of small cell carcinoma of the lung, *Yale J. Biol. Med.,* 54, 187, 1981.
26. Gazdar, A. F., Carney, D. N., Guccion, J. G., and Baylin, S. B., Small cell carcinoma of the lung: cellular origin and relationship to other pulmonary tumors, in *Small Cell Lung Cancer,* Greco, F. A., Oldnam, R. K., and Bunn, P. A., Jr., Eds., Grune & Stratton, New York, 1981, 145.
27. Churg, A., Johnston, W. H., and Stulbarg, M., Small cell squamous and mixed small cell squamous-small cell anaplastic carcinoma of the lung, *Am. J. Surg. Pathol.,* 4, 255, 1980.
28. Saba, S. R., Azar, H. A., Richman, A. V., Solomon, D. A., Spurlock, R. G., Mardelli, I. G., and Kasnic, G., Jr., Dual differentiation in small cell carcinoma (oat cell carcinoma) of the lung, *Ultrastructural Pathol.,* 2, 13, 1981.
29. Goodwin, G., Shaper, J. H., Abeloff, M. D., Mendelsohn, G., and Baylin, S. B., Analysis of cell surface proteins delineates a differentiation pathway linking endocrine and nonendocrine human lung cancers, *Proc. Natl. Acad. Sci. U.S.A.,* 80, 3807, 1983.
30. Goodwin, G. and Baylin, S. B., Relationships between neuroendocrine differentiation and sensitivity to γ-radiation in culture line OH-1 of human small cell lung carcinoma, *Cancer Res.,* 42, 1361, 1982.
31. Baylin, S. B., Jackson, R. H., Lennarz, W., Jr., and Shaper, J. H., The spectrum of human lung cancer cells in culture: a potential model for studying molecular determinants of tumor progression and metastasis, in *Cancer Invasion and Metastasis: Biologic and Therapeutic Aspects,* Nicholson, G. L. and Luca, M., Eds., Raven Press, New York, 1984, 281.
32. Kroener, J. F., Saleh, F., and Howell, S. B., 5-Fu and allopurinol toxicity modulation and phase 2 results in colon cancer, *Cancer Treat. Rep.,* 66, 1133, 1982.
33. Petersen, S. E., Bichel, P., and Lorentzen, M., Flow-cytometric demonstration of tumor-cell subpopulations with different DNA content in human colorectal carcinoma, *Eur. J. Cancer,* 15, 383, 1979.
34. Siracky, J., An approach to the problem of heterogeneity of human tumour-cell populations, *Br. J. Cancer,* 39, 570, 1979.
35. Dexter, D. L., Barbosa, J. A., and Calabresi, P., N,N-Dimethylformamide-induced alteration of cell culture characteristics and loss of tumorigenicity in cultured human colon carcinoma cells, *Cancer Res.,* 39, 1020, 1979.
36. Dexter, D. L., Spremulli, E. N., Fligiel, Z., Barbosa, J. A., Vogel, R., VanVoorhees, A., and Calabresi, P., Heterogeneity of cancer cells from a single human colon carcinoma, *Am. J. Med.,* 71, 949, 1981.

37. Spremulli, E. N., Scott, C., Campbell, D. E., Libbey, N. P., Shochat, D., Gold, D. V., and Dexter, D. L., Characterization of two metastatic subpopulations originating from a single human colon carcinoma, *Cancer Res.*, 43, 3828, 1983.
38. Dexter, D. L., Heterogeneity in human colon cancer, in *Cancer Invasion and Metastasis: Biologic and Therapeutic Aspects,* Nicholson, G. L. and Luca, M., Eds., Raven Press, New York, 1984, 265.
39. MacLean, G. D., Seehafer, J., Shaw, A. R. E., Kieran, M. W., and Longenecker, B. M., Antigenic heterogeneity of human colorectal cancer cell lines analyzed by a panel of monoclonal antibodies. I. Heterogeneous expression of Ia-like and HLA-like antigenic determinants, *J. Natl. Cancer Inst.*, 69, 357, 1982.
40. Alumets, J., Alm, P., Falkmer, S., Hakanson, R., Ljungberg, O., Martensson, H., Sundler, F., and Tibblin, S., Immunohistochemical evidence of peptide hormones in endocrine tumors of the rectum, *Cancer,* 48, 2409, 1981.
41. Cushing, H. and Wolbach, S. B., The transformation of a malignant paravertebral sympathicoblastoma into a benign ganglioneuroma, *Am. J. Path.*, 3, 62, 1927.
42. Kissane, J. M. and Ackerman, L. V., Maturation of sympathetic nervous system, *J. Fac. Radiol. London,* 7, 109, 1955.
43. Visfeldt, J., Transformation of sympathicoblastoma into ganglioneuroma, *Acta Pathol. Microbiol. Scand.*, 58, 414, 1963.
44. Dyke, P. C. and Mulkey, D. A., Maturation of ganglioneuroblastoma to ganglioneuroma, *Cancer,* 20, 1343, 1967.
45. Baylin, S. B., Clonal selection and heterogeneity of human solid neoplasms, in *Design of Models for Testing Cancer Therapeutic Agents,* Fidler, I. J. and White, R. J., Eds., Reinhold, New York, 1981, 50.
46. Baylin, S. B. and Mendelsohn, G., Medullary thyroid carcinoma: a model for the study of human tumor progression and cell heterogeneity, in *Tumor Cell Heterogeneity: Origins and Implications,* Vol. 4, Owens, A. H., Jr., Coffey, D. S., and Baylin, S. B., Eds., Academic Press, New York, 1982, chap. 2.
47. Schimke, R. N. and Hartmann, W. H., Familial amyloid-producing medullary thyroid carcinoma and pheochromocytoma. A distinct genetic entity, *Ann. Intern. Med.*, 63, 1027, 1965.
48. Williams, E. D., Histogenesis of medullary carcinoma of the thyroid, *J. Clin. Pathol.*, 19, 114, 1966.
49. Goster, G. V., MacIntyre, I., and Pearse, A. G. E., Calcitonin production and the mitochondrion-rich cells of the dog thyroid, *Nature (London),* 203, 1029, 1964.
50. Hakanson, R., Owman, C., and Sundler, F., Aromatic L-amino acid decarboxylase in calcitonin-producing cells, *Biochem. Pharmacol.*, 20, 2187, 1971.
51. Pearse, A. G. E., Common cytochemical properties of cells producing polypeptide-hormone secretion with particular reference to calcitonin and the thyroid C cells, *Vet. Rec.*, 79, 587, 1966.
52. Pearse, A. G. E., The cytochemistry and ultrastructure of polypeptide hormone-producing cells of the APUD series and the embryologic, physiologic, and pathologic implications of the concept, *J. Histochem. Cytochem.*, 17, 303, 1969.
53. Beaven, M. A., Doppman, J., Wells, S. A., Jr., and Buja, L. M., Sipple's syndrome; medullary thyroid carcinoma, pheochromocytoma, and parathyroid disease, NIH Conference, *Ann. Intern. Med.*, 78, 561, 1973.
54. Lippman, S. M., Mendelsohn, G., Trump, D. L., Wells, S. A., Jr., and Baylin, S. B., The prognostic and biological significance of cellular heterogeneity in medullary thyroid carcinoma: a study of calcitonin, L-dopa decarboxylase, and histaminase, *J. Clin. Endocrinol. Metab.*, 54, 233, 1982.
55. Van Dyke, D. L., Jackson, C. E., and Babu, V. R., Multiple endocrine neoplasia type (MEN-2) — an autosomal dominant syndrome with a possible chromosome 20 deletion, *Am. Soc. Hum. Gen. Abstr.*, 209, 69A, 1981.
56. Baylin, S. B., Mendelsohn, G., Weisburger, W. R., Gann, D. S., and Eggleston, J. C., Levels of histaminase and L-dopa decarboxylase activity in the transition from G cell hyperplasia to familial medullary thyroid carcinoma, *Cancer,* 44, 1315, 1979.
57. Wells, S. A., Jr., Baylin, S. B., Gann, D. S., Farrell, R. E., Dilley, W. G., Preissig, S. H., Linehan, W. M., and Cooper, C. W., Medullary thyroid carcinoma: relationship of method of diagnosis to pathologic staging, *Ann. Surg.*, 188, 377, 1978.
58. Foulds, L., The experimental study of tumor progression: a review, *Cancer Res.*, 14, 337, 1954.
59. Nowell, P. C., The clonal evolution of tumor cell populations, *Science,* 194, 23, 1976.
60. Dolbeare, F. S and Smith, R. E., Flow cytoenzymology: rapid enzyme analysis of single cells, in *Flow Cytometry and Sorting,* Melamed, M. R., Mullaney, P. F., and Mendelsohn, M. L., Eds., John Wiley & Sons, New York, 1979, 317.
61. Haskill, S., Kivinen, S., Nelson, K., and Fowler, W. C., Detection of intratumor heterogeneity by simultaneous multiparameter flow cytometric analysis with enzyme and DNA markers, *Cancer Res.*, 43, 1003, 1983.

62. Haskill, S., Becker, S., Johnson, T., Marro, D., Propst, R., and Nelson, K., Simultaneous 3 color and electronic cell volume analysis using flow cytometry, in press, 1984.
63. Shapiro, J. R., Yung, W.-K. A., and Shapiro, W. R., Isolation, karyotype, and clonal growth of heterogeneous subpopulations of human malignant gliomas, *Cancer Res.*, 41, 2349, 1981.
64. Yung, A. W., Shapiro, J. R., and Shapiro, W. R., Heterogenous chemosensitivities of subpopulations of human glioma cells in culture, *Cancer Res.*, 42, 992, 1982.
65. Wikstrand, C. J., Bigner, S. H., and Bigner, D. D., Demonstration of complex antigenic heterogeneity in a human glioma cell line and eight derived clones by specific monoclonal antibodies, *Cancer Res.*, 43, 3327, 1983.
66. Bronson, D. L., Andrews, P. W., Solter, D., Cervenka, J., Lange, P. H., and Fraley, E. E., Cell line derived from a metastasis of a human testicular germ cell tumor, *Cancer Res.*, 40, 2500, 1980.
67. Barranco, S. C., Ho, D., Drewinko, B., Romsdahl, M. M., and Humphrey, R. M., Differential responses by human melanoma cells grown *in vitro* to arabinosylcytosine, *Cancer Res.*, 32, 2733, 1972.
68. Barranco, S. C., Drewinko, B., and Humphrey, R. M., Differential responses by human melanoma cells to 1,3-bis-(2-chloroethyl)-1-nitrosourea and bleomycin, *Mutat. Res.*, 19, 227, 1973.
69. Tsuruo, T. and Fidler, I. J., Differences in drug sensitivities among tumor cells from parental tumors, selected variants and spontaneous metastases, *Cancer Res.*, 41, 3058, 1981.
70. Biorklund, A., Hakansson, L., Stenstarn, B., Trope, C., and Ackerman, M., Heterogeneity of non-Hodgkin's lymphomas as regards sensitivity to cytostatic drugs, *Eur. J. Cancer*, 16, 647, 1980.
71. Churg, S. and Ringus, J., Ultrastructural observations on the histogenesis of alveolar rhabdomyosarcoma, *Cancer*, 41, 1355, 1978.
72. Gilbert, F., Balaban, G., Breg, W. R., Gaille, B., Reld, T., and Nichols, W., Homogeneously staining region in a retinoblastoma cell line: relevance to tumor initiation and progression, *J. Natl. Cancer Inst.*, 67, 301, 1981.
73. Fisher, E. R. and Kim, W. S., Primary clear cell thyroid carcinoma with squamous features, *Cancer*, 39, 2497, 1977.

Chapter 3

THE NOWELL HYPOTHESIS AND SUBSEQUENT DEVELOPMENTS

I. INTRODUCTION

The important report by Nowell in 1976 serves as a convenient point to historically divide studies on change and variability within single neoplasms.[1] Nowell hypothesized that genetic instability exists within tumors, and that karyotypic changes in cancer cells produce variant, or mutant cells in a malignancy. These cells, if they survive and proliferate, form distinct clones within the neoplasm. Variant cells would appear throughout the evolution of the tumor, and the development of clonal populations would be a continuous event in the history of a cancer. Intra- and extratumor selection pressures would then select for certain "favored" clones and select against others less suited to survive.[1,2]

The concept to be stressed is that the development of a tumor is a dynamic process.[3] Subpopulations appear and play a role in the evolution of the cancer, which is dependent on the intrinsic properties of each subset of cells and on the extrinsic selection forces impinging of the neoplasm. The Nowell hypothesis, which has gained widespread acceptance, both defines and explains intratumor heterogeneity. Genetic instability in cancer cells causes mutational events that result in the appearance of variant cells in a tumor. This process could begin early in the life of a neoplasm, and continues throughout the growth of the tumor. Subpopulations surviving intra- and extratumor selection pressures coexist in proportions determined by their growth properties, by the nature of interclonal interactions, and by their response to signals from the environment. The tumor thus can become heterogeneous early in its history, and can remain a heterogeneous tissue throughout its development. The degree of variability, the precise mechanisms responsible for generation of mutant cells, and the maintenance of heterogeneity will depend on the particular tumor and host in question. Intraneoplastic diversity has a distinctly individual flavor, a finding that has profound clinical implications.

Another explanation for heterogeneous tumors is that neoplasms are polyclonal in origin.[4] Such a tumor would contain two or more distinct clones from its inception, and diversity would be necessarily an inherent feature of the cancer. There is evidence that some tumors may be of polyclonal origin.[5-6] However, isoenzyme and cytogenetic evidence has clearly demonstrated that the great majority of neoplasms are monoclonal in origin.[3,7] Nowell, in fact, has concluded that most cancers originate each from a single transformed cell, and thus are clonal initially.[1] Therefore, heterogeneity must be generated (in monoclonal tumors) through a process of genetic mutation, as hypothesized by Nowell, or by some other mechanism. However, one can still ask about heterogeneity in polyclonal neoplasms. The multiplicity of phenotypes observed in individual cancers suggests that if the diversity arose in the transformation process itself, a large number of cells must have been transformed at that time. Although one could envision two, three, or even several cells simultaneously sustaining a carcinogenic insult, it is difficult to imagine a large number of cells transformed within a narrow time frame in a tissue. Therefore, one can conclude that if a tumor is polyclonal in origin, the limited diversity existing within the incipient neoplasm must increase further during its evolution due to the same processes responsible for the heterogeneity within tumors of monoclonal origin.

The history of intratumor heterogeneity has thus developed along the following path. Pathologists for decades were aware of variability within both human and animal cancers.[8,9] However, this diversity within individual tumors was felt to be due to host effects on the tumor (extrinsic factors) rather than a result of some intrinsic property of the neoplasm itself. Foulds anticipated the heterogeneous nature of tumors in his classical studies almost thirty years ago.[10-14] However, workers in the areas of tumor biology and experimental therapeutics of cancer were pursuing studies with model systems designed for their simplicity. The key to such a straightforward model as the murine L1210 leukemia was the homogeneity of the cell line.[15,16] The approach taken by many investigators in the fifties and sixties was to work with cancer cell lines that could give reproducible results and would behave in a fairly uniform fashion. The model systems employed were selected for their lack of variability, for an absence of diversity. This approach ignored, for the most part, the phenomenon of tumor progression.

Nowell provided an explanation for the tumor progression described so elegantly by Foulds, and presented the conceptual basis for defining intrinsic tumor heterogeneity. Shortly thereafter, Fidler and Kripke[17] reported that the B16 murine melanoma is heterogeneous for metastatic potential. Cloning of the parent cell line resulted in the isolation of subpopulations, some of which were significantly more effective at lung colonization than others, or the parent line itself. The phenomenology rapidly expanded, and the documentation of distinct subsets of neoplastic cells present in animal and human tumors quickly became extensive. Today the evidence is indeed compelling that most, if not all, solid tumors are heterogeneous. If one looks for differences among neoplastic cells in a cancer, such differences invariably will be found (see Chapters 1, 2, and 5).

Further insights were made on the origins of intraneoplastic diversity following the reports of Nowell, and Fidler and Kripke. Certain predictions can be made concerning the instability of tumor cells, which bear on the biological behavior of neoplasms. If tumor cells are genetically unstable, mutations will result in the appearance of variants. This generation of diversity is responsible for tumor progression, with the evolving cancer characterized by even more instability. Indeed, Harris et al.[18] have calculated that metastatic variants arise in clonal lines of KHT murine sarcoma cell line at a rate of about 10^{-5} per cell per generation. This is higher than would be expected in mammalian cells, where mutation rates are usually 10^{-6} to 10^{-8} per cell per generation. The authors concluded that this represented a dynamic heterogeneity with variants arising at a high rate due to epigenetic or specific genetic mechanisms.

If mutation rates are higher in cancer cells than in normal mammalian cells, this could result in the appearance of variants in a tumor early in its development. One estimate of the early origin of variability in a clone of malignant cells has been provided by a study of transformants isolated from a culture of BALB/3T3 cells.[19] The parent 3T3 line was routinely cultured in agar to check for the appearance of spontaneously transformed cells. One transformed colony was observed, and was removed as soon as it appeared; the cells were grown out in monolayer culture. The transformed cells were reseeded in agar, and five distinct colonies were isolated. These clones were heterogeneous with repect to morphology, clonogenicity in agar, and tumorigenicity in nude mice. Subclones of three of these clones were also obtained, and variability was shown both between and within the subclones. It was concluded that oncogenesis occurred in a single cell, followed by the rapid evolution of diverse phenotypes in the original transformed population.[19] This experiment indicates that diversity in a cancer cell population which was clonal in origin can be quickly generated. Care must be taken in extrapolating the rapidity of events occurring in a transformed 3T3 clone to a human solid tumor that is monoclonal in origin because the two situations are certainly not identical. Nevertheless, a study such as this does provide a lower limit for the time frame required for the genesis of intraclonal diversity in ɔr cell population. Evidence that diversity can also arise early in clonal metastases has also been reported. Talmadge et al. have used spontaneous, clonal B16 metastases to demonstrate that heterogeneity for metastatic potential and drug sensitivity can develop rapidly in these lesions.[20]

It can be hypothesized that the most progressed neoplasms, i.e., those that are the most autonomous, the most aggressive, or the farthest removed from host control, should be comprised of cells that are the most genetically unstable compared to cells in less advanced neoplasms or even to ancestral cells present in that tumor at an earlier time point. Metastases, which are often thought to represent the aggressive extension of a "progressed" cancer, should therefore consist of cells more genetically unstable than cells in the corresponding primary tumor. This hypothesis has been tested experimentally by Cifone and Fidler.[21] These investigators studied cloned cells with differential metastasizing capabilities from each of three murine tumor lines. The spontaneous mutation rates for these subpopulations were determined with respect to two loci, those controlling thiopurine and ouabain resistance, respectively. In each instance, cells from the clone with high metastatic potential had a greater rate of spontaneous mutation at either locus compared to the corresponding clone with low metastatic potential. These results agree with other data demonstrating that irradiating murine tumor cells, which increases their mutation rate, also increases the metastatic potential of the treated cells.[22]

These studies were conducted with cells from murine tumors. Two important and related questions which should be addrssed with human tumors are (1) whether cells responsible for metastases in a patient are more genetically unstable than the (great majority of) cells in the primary tumor, and (2) whether treatment will impact further on this instability.

II. THE DYNAMIC TUMOR

The findings of Foulds, Nowell, and other workers have prompted a consideration of a dynamic tumor model that might contain the following components. The primary tumor itself consists of distinct subpopulations that arose during the evolution of the neoplasm and survived extrinsic and intrinsic selection pressures. These clones would be expected to coexist in a dynamic state where two basic and counteracting forces could interact. One force is the continued genetic instability in neoplastic cells that would result in the appearance of new variants as well as in changes in growth patterns and in susceptibility to selection pressures in existing clones. This evolutionary driving force is responsible for progression in the cancer, resulting in decreased stability, increasing variability, and constant change. An important counteracting force, which limits fluctuations in the neoplasm, is the collection of biologic mechanisms responsible for the maintenance of stability in the heterogeneous tumor.[23-25] Stabilizing interactions among subpopulations in a tumor are discussed in Chapter 7. The net result of the interplay between these two forces would be a tendency towards an equilibrium or a balance between uncontrolled progression on the one hand and a stabilized tumor ecosystem on the other. The extent of progression in a given tumor and the dynamics defining the evolution of that neoplasm will depend on the relative importance of genetic drift vs. factors modulating that drift for a particular cancer.

A completely different type of dynamic tumor model has been proposed over the years from a consideration of kinetic and differentiation events in neoplasms. This model has been described in recent reports by Mackillup et al.[26] and by Selby et al.[27] The essential features of this picture of cancer development include three distinct cell compartments comprised of stem cells, transitional cells, and end cells. Stem cells are defined as clonogenic cells with self-renewal capacity; these cells are the essential targets in therapy. Transitional cells have some clonogenic capability, but do not have the capacity for self-renewal. Transitional cells can add significantly to the tumor burden, and their eradication is critical to debulking, but even a large reduction in the number of these cells will only be palliative. Complete eradication of the stem cell component alone will be curative. End cells in neoplasms are defined as differentiated, nonproliferating progeny of transitional cells.

The picture that emerges is certainly that of a heterogeneous neoplasm but the intraneoplastic diversity occurring in the stem-end cell model is quite different from the more classical model of tumor heterogeneity that has been studied and discussed for the past several years. The description proposed by Mackillup et al. for human solid tumors is patterned after cell renewal in hematopoietic neoplasms, which in turn has been presented as a caricature of self-renewal in normal bone marrow.[28-31] Moreover, marrow stem cells are merely one example of similar processes occurring in various tissues in embryogenesis, organism development, and tissue regeneration.[32,33] In contrast, the appearance and coexistence of a multiplicity of phenotypes in a heterogeneous solid tumor, which is envisioned by many other workers, is not patterned after nor predicted from processes occurring in normal cells and tissues.

We shall refer to tumor evolution as predicted by the Nowell hypothesis as the progression model, and to tumor development as outlined by Selby et al. as the differentiation model. A more complete analysis of the two models produces more questions than it answers. Two cases have to be considered. First, the tumor arises from a transformed primitive stem cell. Second, the tumor develops from a transformed normal nonproliferating adult cell (Figure 1).

The following comments can be made if one considers that the neoplasm arose from a transformed stem cell in a tissue.[34]

At an early stage, the progression and differentiation models can be considered fairly analogous. The tumor is composed of homogeneous, dividing primitive cells with self-renewal capacity. The models diverge at that point in time where heterogeneity is introduced. Maturational events cause heterogeneity in the differentiation scheme, whereas mutational events produce diversity in the progression model. Any maturational change occurring in the sequence of mutational events would be random, not predictable, and thus fortuitous. Differentiation in the stem cell-end cell sequence should be directed, albeit aberrantly, according to the genetic programming in the tissue type from which the tumor arose. Differentiation in the sense that developmental biologists use the term has no *a priori* importance in the progression model, whereas it is a crucial aspect of the stem cell-end cell model.

In the later stages of evolution of a heterogeneous cancer arising from transformed stem cell, one has to consider the mechanisms whereby the diversity is maintained. A tumor can have sizable proportions of subpopulations for any given phenotype, and the means of preservation of the ecosystem in the neoplasm might be expected to be different for each model. In the differentiation scenario, new cells would be continually infused into each compartment via the maturation of stem and transitional cells. The progression sequence would suggest, however, that interactions among subpopulations might be a dominant factor in heterogeneity maintenance.[35] Stem cells in this case would not necessarily play a major role in the (directed) regeneration of variant types.

The following correlate can also be proposed in the situation where the maintenance of heterogeneity breaks down, and one subpopulation ultimately emerges as the dominant subset in the tumor. The neoplasm then would finally become homogeneous and consist of only one population of cells. In the progressed cancer, one would expect an

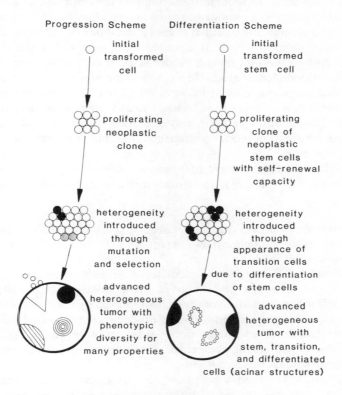

FIGURE 1. The evolution of a heterogeneous neoplasm as might be envisioned from the progression (mutation/selection) model of Nowell or from the stem cell differentiation model of Selby et al. is depicted. The progression scheme has its basis in genetic instability and proliferation of "favored" variant clones (notice that some of the cells are metastasizing). The differentiation scheme is predicated on a programmed differentiation process occurring in a malignant tissue. The two sequelae are simplified views: combinations of these processes as well as other mechanisms may be operative in various neoplasms depending on tumor type. See text for details.

aggressive, autonomous, anaplastic subpopulation of cells, perhaps characterized by a hyperploid and altered complement of chromosomes, to become dominant. In the differentiation model, one could expect that a tumor could become homogeneous only if all the stem and transitional cells differentiated to end cells that were essentially benign, nonproliferative, and mature. The working assumption would be that the progression model would produce ultimately a homogeneous primitive cancer, whereas the differentiation model would result in a homogeneous tumor composed of differentiated, benign tissue.

If carcinogenesis involves an adult, quiescent cell that produces in its early stages a small tumor not very different in its properties from the tissue of origin, a different pattern of evolution would be expected.[36] According to the differentiation model, there should be a temporal process of dedifferentiation, with cancer cells becoming increasingly aggressive, autonomous, and anaplastic. Thus the transit of cells now occurs "down the ladder," from differentiated to transition to stem cells. This is in the opposite direction predicted by the proponents of the stem-end cell model. To further postulate that these stem cells evolving through a dedifferentiation process now proceed back "up the ladder" to become transition and then end cells again is even more difficult to accept. This sort of tumor behavior or dynamics is not particularly appealing conceptually to proponents of either the differentiation or the progression model.

The "predicted" natural history of a cancer that starts with a single transformed adult cell according to the progression model is more defensible. The early tumor is homogenous and well differentiated. Through the processes of mutation and selection, variants arise that are more anaplastic, have a higher ploidy, and are less subject to host control. These more "progressed" clones finally give rise to the primitive, more aggressive cells considered to be the highly malignant (stem) cells of a neoplasm. In this scheme, the progression model is quite suitable, but it is also compatible with a stem cell-derived tumor.

Having traced the natural history of a cancer from either direction (stem cell vs. adult cell transformation), as predicted from either model, the following questions can be asked:

1. Is the differentiation pattern, when it occurs, specific for stem cell tumors only?
2. Is the progression model more consistent with the evolution of a tumor from a transformed adult cell?
3. Does the differentiation pattern occur less frequently than the progression pattern? If so, tumors with a spectrum of stem-transition-end cells might be expected to occur infrequently. This hypothesis is supported by the finding that leukemias and teratorcarcinomas along with a few other cancers that can be clearly defined as differentiated-evolved tumors occur much less frequently than the epithelial solid tumors. However, the presence of primitive (stem) cells in normal colon and breast argues against this hypothesis.
4. As a correlate of (3), are the commonly occurring carcinomas of the lung, colorectum, and breast examples of progression tumors derived from transformed adult cells? Gliomas presumably derived from nondividing cells in brain tissue would be a case in point. Since breast and colon have been reported to contain normal stem cells, the situation with such tissues is less clear. Indeed, one could speculate that some colon cancers are stem cell (crypt cell)-derived, whereas others might be adult cell (tip, or villus cell)-derived.
5. Are the two patterns of tumor evolution mutually exclusive? Or, can one tumor feature both differentiation and progression patterns, either concurrently or at different time points during the development of the neoplasm?

The answers to such queries will come only as more information is obtained about the natural history of solid tumors. Certainly, the conduct of studies that may shed some light on these problems is crucial to our efforts to translate this understanding into efficacious treatment protocols.

There is still another level of complexity in cancers that can be introduced at this point. There seems to be yet another category in which tumors can be placed or defined according to a developing pattern of heterogeneity. Work (see discussion and references in Chapter 2) on heterogeneous small cell lung cancer (SCLC) and medullary thyroid carcinoma (MTC) suggests that a second, distinct stem cell-end cell model might describe these neoplasms. These tumors sometimes progress in patients to produce heterogeneous cancers that signal a poorer prognosis for that patient. In some cases of SCLC, tumor tissue obtained at autopsy contains both SCLC and non-SCLC types; the original tumor appeared to be homogeneously SCLC. In MTC, the biochemical profile of the tumor changes dramatically as the disease progresses; the calcitonin level falls and dopa decarboxylase and diamine oxidase expression is increased. Other changes are also observed.

One interpretation of data from studies on both tumor types is that MTC and SCLC contain stem cells that give rise to other neoplastic cell subpopulations. The evolving heterogeneous lung or thyroid cancer is a more progressed neoplasm arising from programmed changes in the tumor stem cell. This is analogous to a differentiation model, except that the direction of the change is to a higher grade malignancy, not to a differentiated, more benign cancer. The process would be more accurately defined as a controlled dedifferentiation to a more progressed form of the disease. Tumors in this category might be considered as hybrids of those classified clearly as either differentiation model or progression model cancers. SCLC and MTC are both progressing tumors, but they evolve in their malignancy according to a programmed determination and not in an unpredictable, random manner. In this sense, their evolution is reminiscent of that of a teratoma, but the developmental direction taken by the neoplastic stem cell is down the path of dedifferentiation rather than along the maturational vector.

III. MATURATIONAL THERAPY OF HETEROGENEOUS TUMORS

One immediate implication of the defining of a tumor as a differentiation pattern neoplasm rather than as a progression model cancer concerns the sensitivity of a tumor to maturational agents. Stem cell cancers displaying the stem cell-end cell spectrum have been postulated for conceptual reasons to be good targets for differentiation-inducing agents.[37,38] The work of a number of well-known investigators has suggested that leukemias, teratocarcinomas, squamous cell carcinomas, and perhaps melanomas and colon cancers would be good candidates for maturational therapy.[39-42] In these cases, inducing differentiation in the tumor cells could drive these cancer cells to become postmitotic, terminally differentiated benign cells. Therefore, neoplasms falling into the category of the stem-transition-end cell model would be considered logical candidates for treatment with differentiation-inducing compounds.

The intriguing natural histories of small cell lung and medullary thyroid carcinomas suggest that these neoplasms would be exciting targets for maturational agents. Since the SCLC and MTC tumor types combine features of progression model cancers as well as end cell-stem cell neoplasms, it would be fascinating to study the effects of differentiation-inducing chemicals on these tumors.

Tumors classified as progression neoplasms might also be considered as targets for maturational therapy, but for different reasons. Since no differentiation pattern, in the sense developmental biologists use the term, is to be expected in these heterogeneous cancers, there may be no good rationale for driving these cells to end cell status. Indeed, studies done with solid tumor cells using inducers such as butyrate and polar solvents have demonstrated that the effects of such agents on several solid tumor cell types are reversible.[43-45] A postmitotic, terminally differentiated state was not achieved in these cases. However, differentiation-inducing chemicals may be useful in treating these progression model tumors for other reasons (Chapter 8).

First, there is good evidence that inducing chemicals can sensitize cancer cells to other forms of therapy.[46] Polar solvents can enhance the activity of X-irradiation against human and murine carcinoma cells,[47,48] and these compounds can also enhance the acitivity of cis-platinum against human colon cancer cells.[49] Combinations of differentiation-inducing chemicals with other cytotoxic drugs have been suggested. An inducer could enhance the activity of cytotoxic drug by facilitating its transport into tumors cells, by increasing the rate of metabolism of the primary drug to an active form, or by inhibiting enzymatic degradation of the cytotoxic agent. Also, the conversion of estrogen-receptor negative breast cancer cells to estrogen-positive cells by differentiation agents could sensitize these cells to hormone therapy. Therefore, drugs that act as maturational agents against tumors evolving via the stem cell-end cell process could act as a class of biological response modifiers with conventional treatment modalities against progression type tumors.[38,46]

There is another mechanism by which tumors evolving according to the Nowell hypothesis might be affected by maturational agents in a therapeutically favorable manner. The unpredictable and often extensive heterogeneity characterizing solid tumors developing according to the progression model is perhaps the greatest obstacle facing the clinician treating the patient. Differentiation-inducing compounds might cause the conversion of the diverse variant subpopulations to a more common phenotype characterized perhaps by certain differentiated features. It has been hypothesized that this conversion of phenotypes will simplify therapy because fewer bullets will be required to eradicate a tumor that has fewer target subpopulations.[3,15,16,50] In other words, maturational agents may not terminally differentiate progression tumors, but they could "homogenize" them and thus eliminate the ominous treatment implications presented by tumor heterogeneity.

It is reasonable to conclude that the particular pattern of heterogeneity, or the type of heterogeneity, found in a neoplasm as well as among metastatic lesions from that same primary tumor, has an important clinical meaning. The type of heterogeneity encountered may well be crucial in determining the course of therapy to be used. This could be especially important in cases where maturational agents or various biological response modifiers are considered in the therapeutic strategy.

IV. TUMOR HETEROGENEITY AND ONCOGENES

There has been a tremendous interest in the past several years in the role of oncogenes and oncogene products in neoplasia. The retrovirus transforming genes and their homologs (*c-onc* genes) in normal and tumor cells have been reviewed in several reports,[51-54] and are also discussed in Chapter 9. The question can be asked here whether the *c-onc* genes play a role in intratumor heterogeneity.

There has not been sufficient time to explore any but the most rudimentary ramifications of differential oncogene expression in cancer cell subpopulations in heterogeneous tumors and their metastases. However, Buick and Pollak have recently provided some stimulating perspectives on this topic.[55] These workers have proposed the following scheme as an extension of the stem-transition-end cell model to explain the evolution of tumor heterogeneity. There is a clear relationship between the *c-onc* genes and proliferation in cancer cells, embryonic cells and in regenerating rat liver cells.[56-58] The expression of *c-myc* is stimulated during proliferation[56,59] and downregulated when tumor cells differentiate.[60,61] Furthermore, growth factors and their receptors, critical to proliferation of many cell types, are believed to be coded by *c-onc* genes. It has already been established that *erb-b* codes for a product probably homologous to epidermal growth factor receptor and *sis* contains the information for a product almost identical to platelet-derived growth factor.[62,63] Also, these authors speculate that entry into the cell cycle and the capacity for self-renewal (which defines the stem cell) are also deregulated in cancer cells due to either the overexpression of a (onco) gene coding a for a normal protein[64] or to the expression of a (onco) gene coding for an altered protein.[65]

The role of oncogenes in the development of the heterogeneous neoplasm is beginning to emerge as follows. Deviations from normal *c-onc* function could result in an increasing percentage of cells residing in the stem cell compartment. Other derangements in *c-onc* expression will modulate the responses of stem and transition cells to exogenous factors, themselves a product of oncogene sequences. This will result in a variable production of end (differentiated) cells from an unpredictably changing mix of stem and transition cells. The pattern will change for each tumor, but any given tumor will be heterogeneous because of the presence of end, transition, and stem cells, all of which arise due to inappropriate oncogene expression. Buick and Pollak have synthesized the stem cell tumor concept and the oncogene experience to date to provide a molecular explanation for the phenomenon of intraneoplastic diversity.[55] Although work is only beginning in this area, certainly the origins of heterogeneity must lie at least in part in the altered and variable expression in tumor cells of critical DNA sequences (called cellular oncogenes), which function in a highly regulated fashion in normal cells.

V. SUMMARY

In both the progression and the differentiation models, the three-dimensional tumor evolves in time from a single (or possibly a few) transformed cells. The fourth dimension provided by time is a critical feature common to both views. The tumor evolves in a discreet time period, and its composition is constantly changing throughout the duration of its development. Both models are dynamic, and therefore one can only describe the precise heterogeneous makeup of the tumor for a given point in time. At an earlier or later period, the same tumor would have a different composition in terms of number and proportions of subpopulations.

The Nowell hypothesis, which has provided the conceptual underpinning for the phenomenon of tumor progression studied and described so elegantly by Foulds, has genetic instability and appearance of mutants (variants) as its primary tenets. This pattern of tumor development requires new genetic information introduced by mutational events in cancer cells. Accordingly, there is no counterpart in normal development.

The processes occurring in embryogenesis, organogenesis, and organism development require a timely expression of an inherited genetic programming. This includes stem cell proliferation with retention of self renewal capacity, determination of daughter cells with increasing restriction of differentiation options, and finally production of differentiated end cells that perform specific functions in tissues.[66-68] The differentiation model has, as its basis, this pattern of normal tissue development. The tumor is necessarily related to its tissue of origin, and the cancer cells abortively seek to recapitulate the natural history of the cells in the corresponding normal tissue. This perception of the tumor and its development, with an implicit assignment of subsets and functions (i.e., heterogeneity), has been stated by investigators including Van Potter,[69] Sachs,[70] and Pierce.[71]

The differentiation model describes a heterogeneous tumor that is a pathological counterpart to its corresponding normal host tissue. Cancer stem cells strive to attain end cell (differentiated) status. In contrast, the progression held to be responsible for the intratumor and interlesional diversity described recently by so many workers requires a severing of control and increasing autonomy from the tissue of origin. Exciting areas of future research in cancer biology will include the categorizing of basic evolutional patterns operating in tumors, the elucidation of the driving force behind each of these patterns, and the translation of such knowledge into treatment protocols effective against heterogeneous human malignancies.

REFERENCES

1. Nowell, P. C., The clonal evolution of tumor cell populations, *Science* 194, 23, 1976.
2. Nowell, P. C., Tumors as clonal proliferation, *Virchows Arch. B: Cell Pathol.*, 29, 145, 1978.
3. Dexter, D. L. and Calabresi, P., Intraneoplastic diversity, *Biochim. Biophys. Acta*, 695, 97, 1982.
4. Fialkow, P. J., Clonal origin of human tumors, *Biochim. Biophys. Acta*, 458, 283, 1976.
5. Fialkow, P. J., Clonal origin of human tumors, *Annu. Rev. Med.*, 30, 135, 1979.
6. Fialkow, P. J., Clonal and stem cell origin of blood cell neoplasms, *Contemp. Hematol. Oncol.*, 1, 1, 1980.
7. Frei, E., III, Models and the clinical dilemma, in *Tumor Cell Heterogeneity: Origins and Implications*, Vol. 4, Bristol-Meyers Cancer Symposia, Owens, A. H., Coffey, D. S., and Baylin, S. B., Eds., Academic Press, New York, 1982, 181.
8. Dexter, D. L., Kowalski, H. L., Blazer, B. A., Fligiel, Z., Vogel, R., and Heppner, G. H., Heterogeneity of tumor cells from a single mouse mammary tumor, *Cancer Res.*, 38, 3174, 1978.
9. Dunn, T., Morphology of mammary tumors in mice, in *Physiopathology of Cancer*, Homburger, F. and Fishman, N. H., Eds., Paul. B. Hoeber, New York, 1959, 38.
10. Foulds, L., The experimental study of tumor progression: a review, *Cancer Res.*, 14, 337, 1954.
11. Foulds, L., The histologic analysis of mammary tumors of mice, I. Scope of investigations and general principles of analysis, *J. Natl. Cancer Inst.*, 17, 701, 1956.
12. Foulds, L., The histologic analysis of mammary tumors of mice, II. The histology of responsiveness and progression. The origin of tumors, *J. Natl. Cancer Inst.*, 17, 713, 1956.
13. Foulds, L., The histologic analysis of mammary tumors of mice. III. Organoid tumors, *J. Natl. Cancer Inst.*, 17, 755, 1956.
14. Foulds, L., The histologic analysis of mammary tumors of mice. IV. Secretion, *J. Natl. Cancer Inst.*, 17, 783, 1956.
15. Calabresi, P. and Dexter, D. L., Clinical implications of cancer cell heterogeneity, in *Tumor Cell Heterogeneity: Origins and Implications*, Vol. 4, 2nd ed., Owens, A. H., Coffey, D. S., and Baylin, S. B., Eds., Academic Press, New York, 1982, 181.
16. Calabresi, P., Dexter, D. L., and Heppner, G. H., Clinical and pharmacological implications of cancer cell differentiation and heterogeneity, *Biochem. Pharmacol.*, 28, 1933, 1979.
17. Fidler, I. J. and Kripke, M. D., Metastasis results from pre-existing variant cells within a malignant tumor, *Science*, 197, 893, 1977.
18. Harris, J. F., Chambers, A. F., Hill, R. P., and Ling, V., Metastatic variant are generated spontaneously at a high rate in mouse KHT tumor, *Proc. Natl. Acad. Sci. U.S.A.*, 79, 5547, 1982.
19. Rubin, H., Early origin and pervasiveness of cellular heterogeneity in some malignant transformations, *Proc. Natl. Acad. Sci. U.S.A.*, 81, 5121, 1984.
20. Talmadge, J. E., Benedict, K., Madsen, J., and Fidler, I. J., Development of biological diversity and susceptibility to chemotherapy in murine cancer metastases, *Cancer Res.*, 44, 3801, 1984.
21. Cifone, M. A. and Fidler, I. J., Increasing metastatic potential is associated with increasing genetic instability of clones isolated from murine neoplasms, *Proc. Natl. Acad. Sci. U.S.A.*, 78, 6942, 1981.
22. Fisher, M. S. and Cifone, M. A., Enhanced metastatic potential of murine fibrosarcomas treated in vitro with ultraviolet radiation, *Cancer Res.*, 41, 3018, 1981.
23. Miller, B. E., Miller, F. R., Leith, J., and Heppner, G. H., Growth interaction in vivo between tumor subpopulations derived from a single mouse mammary tumor, *Cancer Res.*, 40, 3977, 1980.
24. Miller, B. E., Miller, F. R., and Heppner, G. H., Interactions between tumor subpopulations affecting their sensitivity to the antineoplastic agents of cyclophosphamide and methotrexate, *Cancer Res.*, 41, 4378, 1981.
25. Poste, G., Doll, J., and Fidler, I. J., Interactions among clonal subpopulations affect stability of the metastatic phenotype in polyclonal populations of B16 melanoma cells, *Proc. Natl. Acad. Sci. U.S.A.*, 78, 6226, 1981.
26. Mackillup, W. J., Ciampi, A., Till, J. E., and Buick, R. N., A stem cell model of human tumor growth: implications for tumor cell clonogenic assays, *J. Natl. Cancer Inst.*, 70, 9, 1983.
27. Selby, P., Buick, R. N., and Tannock, I., A critical appraisal of the "human tumor stem-cell assay", *N. Eng. J. Med.*, 308, 129, 1983.
28. Pluznik, D. H. and Sachs, L., The cloning of normal "mast" cells in tissue culture, *J. Cell Comp. Physiol.*, 66, 319, 1965.
29. Bradley, T. R. and Metcalf, D., The growth of mouse bone marrow cells in vitro, *Aust. J. Biol. Med. Sci.*, 44, 287, 1966.
30. Senn, J. S., McCulloch, E. A., and Till, J. E., Comparison of the colony forming ability of normal and leukemic human marrow in cell culture, *Lancet*, 2, 597, 1967.
31. Robinson, W. A., Kurnick, J. E., and Pike, B. L., Colony growth of human leukemic peripheral blood cells in vitro, *Blood*, 38, 500, 1971.

32. Lajtha, L. G., Stem cell concepts, *Differentiation*, 14, 23, 1979.
33. Hall, A. K., A stem cell is a stem cell is a stem cell, *Cell*, 33, 11, 1983.
34. Pierce, G. B., Nakane, P. K., Hernandez-Martinez, A., and Ward, J. M., Ultrastructural comparison of differentiation of stem cells of murine adenocarcinomas of colon and breast with their normal counterparts, *J. Natl. Cancer Inst.*, 58, 1329, 1977.
35. Heppner, G. and Miller, B. E., Biological variability of mouse mammary neoplasms, in *Design of Models for Testing Cancer Therapeutic Agents*, Fidler, I. J. and White, R. J., Eds., Van Nostrand Reinhold Co., New York, 1982, chap. 4.
36. Uriel, J., Cancer, retrodifferentiation, and the myth of Faust, *Cancer Res.*, 36, 4269, 1976.
37. Sachs, L., Control of normal cell differentiation and the phenotypic reversion of malignancy in myeloid leukemia, *Nature (London)*, 274, 535, 1978.
38. Dexter, D. and Calabresi, P., Cancer cell differentiation, in *Pancreatic Tumors in Children*, Humphrey, G. B., Eds., Martinus Nijhoff, The Hague, 1982.
39. Stevens, L. C., The biology of teratomas, *Adv. Morphog.*, 6, 1, 1967.
40. Collins, S. J., Ruscetti, F. W., Gallagher, R. E., and Gallo, R. C., Terminal differentiation of human promyelocytic leukemia cells induced by dimethyl sulfoxide and other polar compounds, *Proc. Natl. Acad. Sci., U.S.A.*, 75, 2458, 1978.
41. Huberman, E., Heckman, C., and Langenbach, R., Stimulation of differentiated functions in human melanoma cells by tumor-promoting agents and dimethyl sulfoxide, *Cancer Res.*, 39, 2618, 1979.
42. Dexter, D. L., Barbosa, J. A., and Calabresi, P., N,N-Dimethylformamide-induced alteration of cell culture characteristics and loss of tumorigenicity in cultured human colon carcinoma cells, *Cancer Res.*, 39, 1020, 1979.
43. Dexter, D. L. and Hagar, J. C., Maturation-induction of tumor cells using a human colon carcinoma model, *Cancer*, 45, 1178, 1980.
44. Kim, Y. S., Tsao, D., Siddiqui, B., Whitehead, J. S., Arnstein, P., Bennett, J. J., and Hicks, J., Effects of sodium butyrate and dimethylsulfoxide on biochemical properties of human colon cancer cells, *Cancer*, 45, 1185, 1980.
45. Dexter, D. L., Konieczny, S. F., Lawrence, J. B., Shaffer, M., Mitchell, P., and Coleman, J. R., Induction by butyrate of differentiated properties in cloned murine rhabdomyosarcoma cells, *Differentiation*, 18, 115, 1981.
46. Dexter, D. L., Leith, J. T., Crabtree, G. W., Parks, R. E., Jr., Glicksman, A. S., and Calabresi, P., N,N-Dimethylformamide-induced modulation of responses of tumor cells to conventional anticancer treatment modalities, in *Maturation Factors and Cancer*, Moore, M. A. S., Ed., Raven Press, New York, 1982.
47. Leith, J. T., Gaskins, L. A., Dexter, D. L., Calabresi, P., and Glicksman, A. S., Alteration of the survival response of two human colon carcinoma subpopulations to x-irradiation by N,N-dimethylformamide, *Cancer Res.*, 42, 30, 1982.
48. Leith, J. T., Brenner, H. J., DeWyngaert, J. K., Dexter, D. L., Calabresi, P., and Glicksman, A. S., Selective modification of the x-ray survival response of two mouse mammary adenocarcinoma sublines by N,N-dimethylformamide, *Int. J. Radiat. Oncol. Biol. Phys.*, 7, 943, 1981.
49. Dexter, D. L., DeFusco, D. J., McCarthy, K., and Calabresi, P., Polar solvents increase the sensitivity of cultured human colon cancer cells to cis-platinum and mitomycin-C, *Proc. Am. Assoc. Cancer Res.*, 24, 267, 1983.
50. Dexter, D. L., Heterogeneity in human colon cancer, in *Cancer Invasion and Metastasis: Biological and Therapeutic Aspects*, Nicolson, G. L. and Milas, L., Eds., Raven Press, New York, 1984, 265.
51. Land, H., Parada, H. F., and Weinberg, R. A., Cellular oncogenes and multistep carcinogenesis, *Science*, 222, 771, 1983.
52. Bishop, J. M., Retroviruses and cancer genes, *Adv. Cancer Res.*, 37, 1, 1982.
53. Duesberg, P. H., Retroviral transforming genes in normal cells?, *Nature (London)*, 304, 219, 1983.
54. Bishop, J. M., Cellular oncogenes and retroviruses, *Annu. Rev. Biochem.*, 52, 301, 1983.
55. Buick, R. N. and Pollak, M. N., Perspectives on clonogenic tumor cells, stem cells, and oncogenes, *Cancer Res.*, 44, 4909, 1984.
56. Campisi, J., Gray, H. E., Pardee, A. B., Dean, M., and Sonenshein, G. F., Cell-cycle control of c-*myc* but not c-*ras* expression is lost following chemical transformation, *Cell*, 36, 241, 1984.
57. Mullar, R., Slamon, D. J., Tremblay, J. M., Cline, M. J., and Verma, I. M., Differential expression of cellular oncogenes during pre- and postnatal development of the mouse, *Nature (London)*, 299, 640, 1982.
58. Goyette, M., Petropoulos, C. J., Shank, P. R., and Fausto, N., Expression of a cellular oncogene during liver regeneration, *Science*, 219, 510, 1983.
59. Kelly, K., Cochran, B. H., Stiles, C. D., and Leder, P., Cell cycle specific regulation of the c-*myc* gene by lymphocyte mitogens and platelet-derived growth factor, *Cell*, 35, 603, 1983.

60. Reitsma, P. H., Rothberg, P. G., Astrin, S. M., Trial, J., Bar-Shavit, Z., Hall, A., Teitelbaum, S. L., and Kahn, A., Regulation of *myc* gene expression in HL60 leukemia cells by a Vitamin D metabolite, *Nature (London)*, 304, 602, 1983.
61. Grosso, L. E. and Pitot, H. D., Modulation of *C-myc* expression in the HL-60 cell line, *Biochem. Biophys. Res. Commun.*, 119, 473, 1984.
62. Downward, J., Yarden, Y., Mayes, E., Scrace, G., Totty, N., Stockwell, P., Ullrich, A., Schlessinger, J., and Waterfield, M. D., Close similarity of epidermal growth factor receptor and *V-erb-B* oncogene protein sequences, *Nature (London)*, 307, 521, 1984.
63. Waterfield, M. D., Scrace, G. T., Whittle, N., Stoobant, P., Johnsson, A., Wasteson, A., Westermark, B., Heldin, C. H., Huang, J. S., and Denel, T. F., Platelet-derived growth factor is structurally related to putative transforming protein p28^{sis} of simian sarcoma virus, *Nature (London)*, 304, 35, 1983.
64. Chang, E. H., Furth, M. E., Scolnick, E. M., and Lowy, D. R., Tumorigenic transformation of mammalian cells induced by a normal human gene homologous to the oncogene of the Harvey murine sarcoma virus, *Nature (London)*, 297, 479, 1982.
65. Reddy, E. P., Reynolds, R. K., Santos, E., and Barbacid, M., A point mutation is responsible for the acquisition of transforming properties by the T24 human bladder carcinoma oncogene, *Nature (London)*, 300, 149, 1982.
66. Bennett, D., Boyse, E. A., and Old, L. J., Cell surface immunogenetics in the study of morphogenesis, in *Le petit Symposium, London,* Silvestri, L. G., Ed., Elsevier, Amsterdam, 1971, 247.
67. Holtzer, H., Cell lineages, stem cells and quantal cell cycle concept, in *Stem Cell and Tissue Homeostasis*, Lord, B. I., Potten, C. S., and Cole, R. J., Eds., Cambridge Univ. Press, London, 1977, 1.
68. Gurdon, J. B. and Woodland, H. R., On the long term control of nuclear activity during cell differentiation, *Curr. Top. Devel. Biol.*, 5, 39, 1970.
69. Potter, V. R., Phenotypic diversity in experimental hepatomas: the concept of partially blocked ontogeny, *Br. J. Cancer*, 38, 1, 1978.
70. Sachs, L., The differentiation of myeloid leukemia cells: new possibilities for therapy, *Br. J. Haematol.*, 40, 509, 1978.
71. Pierce, G. B., The benign cells of malignant tumors, in *Developmental Aspects of Carcinogenesis and Immunity,* King, T. J., Ed., Academic Press, New York, 1974, 3.

Chapter 4

METASTASIS

"As to the nature of the material which passes from the tumour, it must be either something dissolved in the juices, or else solid particles of some kind. Looking to a secondary cancerous infection of the peritoneum, we can hardly escape the conclusion that it is finely divided solid. The secondary tumours are not regularly distributed over the peritoneal surface but occur here and there or in groups, just as if solid particles had been carried and produced their effects where they got leave to lie. Then also it must be solid particles which are arrested by the lymphatic glands and give rise to the secondary tumours there, as it were by embolism. The probability is that the actual cells of the tumours were carried off and deposited at a distance."
J. Coats, 1883.[1]

I. INTRODUCTION

The dissemination of tumor cells from a primary neoplasm to distant sites with subsequent colonization of various organs is the crucial event in the natural history of the disease in most cancer patients. Sugarbaker's statement[2] that "metastasis continues to be the most devastating event for the patient with an established primary tumor" summarizes a universally accepted finding: the major problem in cancer is metastasis, not the primary neoplasm. An obvious corollary to this statement is that progress in controlling cancer will depend, to a great extent, on our ability to control metastatic disease. The major reason why clinicians have been unsuccessful in curing many patients with solid tumors is that there have been no effective treatments against the secondary tumors in these individuals. Most of these patients are not dying of their primary tumors; they are succumbing to disseminated disease.

Much has been written about metastasis. Indeed, from the time of Galen investigators have attempted to understand and explain the dissemination process. Sugarbaker, in his excellent review of metastasis,[2] has provided a good summary of the history of work in this field. A humoral concept of tumor spread prevailed until late in the 19th century. Even famous physicians of that period, including Virchow and Paget, believed that some humoral substance (fluid) produced by cells in a primary tumor caused secondary tumor deposits.[3,4] Finally, work by Thierach, Langenback, and Waldeyer, among others, established the cellular basis of metastasis.[2,5-7]

The clue to this insidious emigration phenomenon lies in the properties of the cells within the primary cancer. The progression in solid tumors, resulting in the appearance of variant subsets of cells, would seem to have as its end point the production of certain aggressive cells that have a propensity to colonize. It can be hypothesized that the ultimate goal of the heterogeneous primary neoplasm is the formation and exportation of specialized colonizers who bring about the demise of the host.

This chapter is not meant to serve as a review of the field of tumor cell dissemination. Rather, we shall attempt to briefly review the metastatic process, then examine the role of heterogeneity in metastasis, and finally discuss the clonality of metastases. Approaches to the treatment of secondary tumors in patients based on information from preclinical studies and clinical experience with metastases will be presented in Chapter 9. The existence within heterogeneous tumors of subpopulations of cancer cells that are functionally specialized for colonization provides an important theme. The focus within the context of that theme will be on the human disease as much as possible. However, much of what we know about the dissemination process itself and about the properties of those cells with the propensity to metastasize has come from work with animal tumors. These studies must be considered as well. The reader should

Table 1
EVENTS IN METASTASIS

Growth of primary neoplasm
Active invasion of surrounding tissues
Mechanical dislodgement of tumor cells
Entry into blood and lymph vessels; embolization
Survival in circulation
Interactions possible with neoplastic and normal cells in circulation
Tumor cell arrest by host tissues
Extravasation
Establishment of vasculature
Proliferation in new site

Note: Summary of the events or "steps" in the metastatic process. A tumor cell must complete all these steps in order to clone a metastasis. This list should be considered only as a guide to the reader, as other orderings or subdivisions of events are certainly possible.

realize that even with the extensive clinical literature available on the nautural history of cancers in patients and the large number of experimental studies reported over the years, there is still much that we do not know or understand about metastasis. Fresh approaches from innovative scientists and clinicians will be necessary to make an impact on the management of patients with disseminated disease.

II. THE METASTATIC PROCESS

The sequelae of events in metastasis constitutes one of the most grimly fascinating occurrences in nature. The process is both scientifically engrossing and depressing to study; in spite of all we have learned about tumor dissemination, metastatic disease is untreatable in many patients. There have been a number of excellent recent reviews on metastasis.[2,8-11]

Fidler, and Sugarbaker have referred to the cell that completes the steps in its colonization journey and initiates a new tumor deposit as a "decathalon winner".[2,12] The steps in metastasis can be outlined as follows (Table 1). Tumor cells must leave the primary neoplasm and release into the circulation. A high pressure gradient of interstitial fluid from the tumor center to its outer perimeters probably helps flush tumor cells to propitious locations for egress into surrounding tissue and blood or lymphatic channels.[13-16] There have been extensive studies, particularly by Liotta and colleagues, on the presence and function of proteolytic enzymes in tumors.[17-21] The degradative action of the various types of collagenases and elastase on the basement membrane of juxtaposing normal tissues is believed to be important in aiding the neoplastic cells to create points of access to the circulation. Finally, the locomotor ability of the tumor cells helps them maneuver into the blood stream.[22-24]

The properties of cells in the primary tumor outlined above are often referred to as parameters of tumor cell invasion, and invasion is often defined as the first step in the metastatic process. Many laboratories are exploring the invasive properties of tumor cells and possible methods of inhibiting this crucial early event in metastasis. Following successful entry into the blood or lymphatic circulation, the tumor cell can begin its long journey to a distant organ.

The sojourn of the migrating cancer cell though the circulation is extremely perilous for the cell, and great numbers of the tumor cells do not survive the experience.[2,10] This is true whether the migration is hematogenous or occurs through lymphatic channels.

Studies with some animal tumor models have demonstrated that there is significant cancer cell destruction in the lymph nodes; these findings have led to the filter barrier concept of lymphatic tumor spread.[25,26] Although this concept is controversial, as evidence also exists that indicates that neoplastic cells can pass freely through lymph nodes,[27-29] there is reason to believe that lymph nodes can serve as obstacles to the spread of some cancers.

The data on hematogenous metastases are more clear. The blood circulation presents a severe, hostile environment to neoplastic cells. Tumor cells from solid tissues are not well equipped to survive in suspension, which they must do while circulating in the blood.[8] Cells undergoing hematogenous dissemination are also susceptible to mechanical damage. Furthermore, natural killer cells, lymphocytes or macrophages may produce an immune response against tumor cells in the blood stream.[30-32] Sugarbaker has pointed out that only a small proportion of cancer cells from various solid tumor models that are injected i.v. survive to form metastases.[2] In summary, the blood stream constitutes an alien and destructive environment for cancer cells derived from solid tumors.

Those few cells that survive their transit through the circulation encounter capillary beds, where a further series of events is necessary before a metastasis is cloned. It has been postulated that aggregation with platelets or other cancer cells is critical to the tumor cell's survival in the blood stream and that clumps of tumor cells have an advantage once they reach capillary vessels.[33,34] The cancer cell is now poised to colonize the new site, and a favorable combination of neoplastic cell properties and certain elements of capillary physiology is thought to result in ultimate implantation of the tumor cell(s) in the target organ. These factors could include tumor cell affinity for target organ tissue (selective "stickiness")[35,36] and a flexible cytoskeleton ("deformability")[37,38] as well as the exposure of tissue basement membrane to cancer cells due to contraction or shedding of endothelial cells.[39,40]

A cell that has managed to complete this rather formidable sequence of steps may then begin to proliferate and form a secondary tumor deposit if it can continue to evade the host immune response, if it can establish its own vascularization,[41] and if it possesses favorable growth kinetics.[42] The incipient tumor nidus is able to induce the host to provide a vascular system to ensure continued growth of the neoplasm via a tumor angiogenesis factor. This factor has been extensively studied by Folkman and colleagues,[43-45] and is another example of the tumor's ability to exploit the host to its own advantage.

This section is not intended as a comprehensive review of the process of metastasis, but only as an outline of the events involved. Much has been written on the subject (for review, see References 2, and 8 through 11), and certainly not everything has been included here. Rather, it is hoped that this summary of the steps in dissemination will serve as a helpful introduction to the subsequent discussion of the importance of tumor heterogeneity to this process.

III. METASTASIS — A RANDOM OR SELECTIVE PROCESS?

No one has disputed the clinical finding that metastasis is actually a rare event. Since a 1-cm^3 tumor weighing 1 g contains approximately 10^9 cells, patients with a tumor burden of tens or hundreds of grams would be expected to develop hundreds or thousands of metastases if dissemination and successful colonization at distant sites occurred at a frequency of even 10^{-6}. The great majority of patients with metastatic disease never present with thousands or even hundreds of secondary tumor deposits.[46,47] Clearly, the process of metastasis is quite inefficient, and it is only the rare cancer cell that produces a secondary tumor.[10,11,48,49] What is not clear from studies of patients or animals bearing tumors is whether all cells in a tumor have an equal, albeit low probability of disseminating, or whether a few cells within a tumor are intrinsically more capable of producing a metastasis.

This question is central in assessing the relevance of tumor heterogeneity to the natural history and progression of the disease in patients. It is also crucial to our understanding of what we are facing in our attempts to treat individuals with metastatic disease. If heterogeneity for metastatic capacity (1) exists, and (2) is linked with heterogeneity for response to treatment modalities, diversity within neoplasms will have presented us with the worst of all worlds.

It is important to understand that we are addressing the question of intrinsic intratumor heterogeneity for metastatic capability vs. metastasis as a stochastic event in a neoplasm. It has been appreciated since the time of Coats,[50] Paget,[51] and other[52,53] that the target organ or "soil" can select for colonizing cancer cells. The existence of "extrinsic selection" by distant organs is not questioned today, and the clinical literature is replete with examples of different tumors metastasizing to favored sites. Breast cancer spreads to lung, liver, and bone.[54] Colorectal cancer disseminates to liver and to lung; pancreatic and gastric cancers frequently metastasize to liver.[55,56] Bone involvement occurs in about 80% of patients with prostatic carcinoma.[57] The list could be extended for many tumor types, and clearly there is organ preference, or "organ selection" for disseminating tumor cells. However, have these cells been previously selected in the primary tumor for metastatic potential? Is there intraneoplastic diversity for the metastatic phenotype?

Early work with murine neoplasms that gradually converted from a solid tumor form to an ascitic form suggested that selection was occurring within the tumor population for a more aggressive subset of cells.[58,59] The definitive study was published by Fidler and Kripke in 1977.[60] These investigators injected cells from clones of B16 isolated from a parent culture and also parent cells into recipient syngeneic mice. They demonstrated that the clones differed markedly from each other and from the parent line in their respective abilities to colonize the lungs of the host mice. Application of the fluctuation assay devised originally by Luria and Delbruck[61] confirmed that selection and not adaptation was responsible for the variability among the clones. Finally, subcloning experiments provided evidence that the results were not due to the cloning technique itself. Similar results were also obtained with a murine tumor of recent origin,[62] and from more extensive work with spontaneous metastases from murine fibrosarcomas.[63] These studies definitively asnwered the question of whether metastasis results from a dynamic selection process or from a random, accidental "spreading" of tumor cells. There is intrinsic heterogeneity for the metastatic phenotype, and this finding has had a great impact on how we view tumor dissemination and on how we would plan to combat it.

One can probe a little more deeply into the metastatic phenotype and ask whether cells with dissemination potential are also more invasive, or do they "follow" the invasive cells into the circulation. An interesting relationship between the metastatic phenotype on one hand[64] and the invasive phenotype on the other[65] is beginning to emerge, at least for certain murine tumors. Starkey et al. have used bovine lens capsule basement membrane and three pairs of tumor cell variant lines with differential metastatic capacity to address this question.[66] These lines included the +SA and -SA mouse mammary tumor lines, the RT7-4bs and RT7-4b-Ls rat hepatocarcinoma cell lines, and the F10 and F1 B16 mouse melanoma clones. For each pair, the more metastatic line was more effective in digesting the lens capsule basement membrane, suggesting that proteolysis-mediated invasive events may be important in the emigration of metastasizing subpopulations contained in heterogeneous solid tumors. The role of proteases in the metastatic process was also implicated in a recent study with a rat mammary tumor cell line.[67] Subpopulations of the 13762 mammary adenocarcinoma line were obtained by dilution cloning, and were assessed for plasminogen activator levels and for the ability to produce lung metastases via tail vein injection. A strong correlation was found between plasminogen activator expression and metastatic potential. There is little or no information currently available on whether cells from human cancers which display the metastatic phenotype also possess invasive properties. Obviously much work remains to be done in this area using human solid tumors.

IV. ARE METASTASES CLONAL?

There is yet another unknown in the tumor dissemination equation which we should address before we can begin to consider the implications of selected cell metastasis in patients. The impact of metastasis on the detection and therapy of cancers will be covered in a later chapter. However, one aspect of tumor dissemination that should be discussed in a chapter focusing on the biology of the process itself is the question of the clonality of metastases. This problem should be viewed particularly from the perspective of whether metastases that are each monoclonal, or homogeneous, would confer a different prognosis on the patient compared to secondary tumors that are polyclonal, or heterogeneous.

There are several fundamental issues at stake here. First, do single cells seed organs and clone metastases, or do secondary tumor deposits arise from clumps of tumor cells? Further, if metastases arise from tumor emboli, are these aggregates each homogeneous or heterogeneous collections of neoplastic cells? Second, where clonal, homogeneous metastases occur in a host, can they still be distinct from one another; does interlesional heterogeneity exist among intralesionally homogeneous metastases? Third, whether metastases are monoclonal (this is usually referred to as clonal) or bi- or even polyclonal, does the new tumor nidus eventually recapitulate some or all of the heterogeneity present in the primary tumor?[49,68]

We have already reviewed evidence that circulating cancer cell clumps may have an advantage in initiating secondary tumor deposits. It would still be possible for a tumor embolus to develop a homogeneous metastasis if (1) only one cell in the aggregate gave rise to the new tumor or (2) if the embolus were homogeneous in its composition to start with. There is some indication from experimental work on murine neoplasms and human tumor tissue material that zonal heterogeneity exists in some cancers.[69] Discreet geographical positioning of subpopulations within a tumor should facilitate the dislodging of clumps of sister cells into the circulation to produce homogeneous emboli capable of cloning a homogeneous metastasis. It might be expected that zonal microheterogeneity would be the prevailing pattern in many solid tumors if (1) clones formed in nests that were not easily penetrated by other clones and (2) there was not a great deal of instability within clones causing significant intraclonal variation. If, however, an embolus is formed by aggregation of single tumor cells in the circulation, or by the merging of small clumps of tumor cells, each homogeneous yet distinct from the other, or by the sticking of single cells to preformed clumps, it will be heterogeneous by the time it reaches the target organ. It would seem reasonable that any of the above could occur depending on the primary tumor type, size, growth kinetics, the influence of host factors, etc. (Figure 1).

In summary, a metastasis could be clonal if it developed from a single cell, or from a homogeneous tumor embolus. A heterogeneous clump of cancer cells would be more likely to develop a multiclonal secondary tumor unless only one cell was clonogenic.[2] There have been several studies that shed some light on this puzzle, which can be

INTRALESIONAL AND INTERLESIONAL HETEROGENEITY

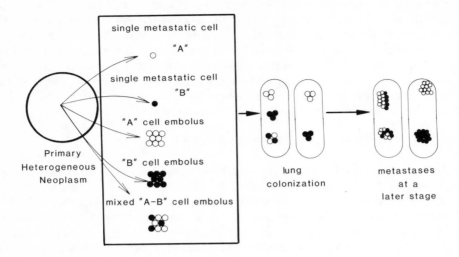

FIGURE 1. Schematic drawing of pulmonary metastases from hematogenous dissemination of two metastatic cell subpopulations from a heterogeneous primary neoplasm. Cell clone "A" (○) and cell clone "B" (●) can travel in the circulation as either single cells, homogeneous emboli, or heterogeneous emboli. The heterogeneous emboli could result from "mixed" clumps emigrating from the primary cancer or from aggregation of "A" and "B" single cells or homogeneous emboli in the blood stream. Initially, depending on the type of the seeding cell or embolus, micrometastases can be heterogeneous or homogeneous. With time, diversity could develop in (originally) homogeneous secondary lesions via the same mechanisms responsible for the appearance of heterogeneity in the primary tumor.

summarized as follows. Talmadge et al.[68] introduced chromosome breaks and rearrangements in K-1735 murine melanoma cells by X-irradiation. Mice were inoculated with cells in the footpad, and lung metastases were recovered from mice after an appropriate time period and analyzed. Most of the secondary tumors exhibited distinct patterns of the chromosomal markers. The authors concluded that many of the lung metastases were clonal in origin, and that they had probably each developed from different progenitor cells.[68] Reeve and Twentyman examined spontaneous (from solid tumors) or artificial (from i.v. injection) metastases developing from an inoculum of murine sarcoma RIF-1 cells, and found that metastasis was a clonal occurrence in this system.[70] The parent RIF-1 line contains diploid and tetraploid subpopulations; metastases were characterized by a single level of ploidy. Spremulli et al. have also used ploidy differences between subpopulations isolated from a heterogeneous human colon tumor line to demonstrate that hematogenous metastases resulting from an s.c. tumor composed of a mixture of the two subpopulations were clonal. The metastases contained only cells from the hyperploid clone.[71]

Other studies with B16 melanoma model have been reported that indicate that metastases can be of either monoclonal or polyclonal origin. Poste et al. analyzed a series of artificially (from i.v. inoculation) produced lung metastases for two phenotypes: metastatic potential and drug resistance.[72] The great majority (18 of 22) of individual lung metastases were homogeneous for both phenotypes. This intralesional clonal homogeneity is evidence of monoclonal origin for these secondary tumor. However, different results were obtained when spontaneous lung metastases (from s.c. tumors) were examined. Four of 18 lung tumor foci were defined as of polyclonal origin because each was heterogeneous for drug markers. Six other metastases were intraclonally heterogeneous for metastatic potential and were judged to most likely be of polyclonal origin. The authors commented that the intraclonal metastatic heterogeneity demonstrated in both spontaneous and (a few) artificial metastases could have resulted from the emergence of variants during the development of a secondary tumor that was monoclonal at its inception. This could not be the case, however, in a metastasis heterogeneous for drug resistance markers, as the experimental design utilized mixtures of clones with different drug markers to inoculate the mice. Therefore, the authors concluded that a least four of the metastases were truly polyclonal in origin.[72]

A review of the relatively few laboratory studies that have addressed the issue of clonality of metastases thus provides evidence that metastases can be monoclonal in origin, but polyclonal secondary tumors also occur. It would seem that intralesional homogeneity (and intralesional heterogeneity) would be dependent on many factors including tumor type, method of introducing tumor cells into the host (i.e., location, route of injection), the organs from which the metastases were obtained, the time allowed for the metastasis to develop, etc. It must also be remembered that interlesional heterogeneity can exist among metastases that each exhibit intralesional homogeneity. Certainly, many studies with murine neoplasms and human tumor xenografts need to be performed, and clinical material (biopsies of metastases and cells from ascitic fluids and pleural effusions) should be analyzed. However, considering the acknowledged existence and postulated importance of tumor cell emboli in the metastatic process (which favors a polyclonal origin for secondary tumors), it is intriguing that several studies have demonstrated that some metastases are monoclonal in origin.

V. SUMMARY

The biology of the metastatic process is extremely complex and still incompletely understood. The finding only a few years ago that selected cells in primary neoplasms have an increased capability to disseminate focused attention on the existence of the metastatic phenotype within heterogeneous solid tumors. It also created pessimism among investigators because it indicated that interlesional heterogeneity could be expected within the host (or the patient). This has been shown to be the case even when metastases in an animal are monoclonal, as they can be each derived from different progenitor cells. The extent of the impact of this grim finding depends on the demonstration that there is interlesional heterogeneity for the treatment sensitivity phenotype (a collection of modality response phenotypes). This is a topic of extreme importance, and it will be discussed in Chapters 8 and 9 which are devoted to the responses of heterogeneous primary tumors and their metastases to various therapies in both the experimental and the clinical setting.

REFERENCES

1. Coats, J., Ed., *A Manual of Pathology*, Henry Lee & Sons Co., Philadelphia, 1883, 177.
2. Sugarbaker, E. V., Cancer metastasis: a product of tumor-host interactions, *Curr. Prob. Cancer*, 7, 3, 1979.
3. Virchow, R., *Cellular Pathology*, (translated by F. Chance), Lippincott, Philadelphia, 1863, 219.
4. Paget, J., *Lectures on Surgical Pathology*, delivered at the Royal College of Surgeons of England, Longmans, Green & Co., London, 1863, 580.
5. Langenbeck, H., On the development of cancer in the veins, and the transmission of cancer from man to lower animals, *Edinburgh Med. Surg. J.*, 55, 251, 1841.

6. Onigbo, W. I. B., A history of hematogenous metastasis, *Cancer Res.*, 30, 2821, 1970.
7. Wilder, R. J., The historical development of the concept of metastasis, *J. Mt. Sinai Hosp.*, 23, 728, 1956.
8. Weiss, L., A pathobiologic overview of metastasis, *Semin. Oncol.*, 4, 5, 1977.
9. Roos, E. and Dingemans, K. P., Mechanisms of metastasis, *Biochim. Biophys. Acta,* 560, 135, 1979.
10. Hart, I. R. and Fidler, I. J., The implications of tumor heterogeneity for studies on the biology and therapy of cancer metastasis, *Biochim. Biophys. Acta,* 651, 37, 1981.
11. Nicolson, G. L., Cancer metastasis: organ colonization and the cell-surface properties of malignant cells, *Biochim. Biophys. Acta,* 695, 113, 1982.
12. Fidler, I. F., Tumor heterogeneity and the biology of cancer invasion and metastasis, *Cancer Res.*, 38, 2651, 1978.
13. Swabb, E. A., Wei, J., and Gullino, P. M., Diffusion and convection in normal and neoplastic tissues, *Cancer Res.*, 34, 2814, 1974.
14. Lunscken, C. and Strauli, P., Penetration of an ascitic reticulum cell sarcoma of the golden hamster into the body wall and through the diaphragm, *Virchows Arch B: (Cell. Pathol.)*, 17, 247, 1975.
15. Butler, T. P. and Gullino, P. M., Bulk transfer of fluid in the interstitial compartment of mammary tumors, *Cancer Res.*, 35, 3084, 1975.
16. Gullino, P. M., *In vivo* release of neoplastic cells by mammary tumors, *Gann,* 20, 49, 1977.
17. Dresden, M. H., Heilman, S. A., and Schmidt, J., Collagenolytic enzymes in human neoplasms, *Cancer Res.*, 32, 993, 1972.
18. Liotta, L. A., Kleinerman, J., Catanzaro, P., and Rynbrandt, D., Degradation of basement membrane by murine tumor cells, *J. Natl. Cancer Inst.*, 58, 1427, 1977.
19. Liotta, L. A., Tryggvason, K., Garbison, S., Hart, I., Foltz, C. M., and Slafie, A., Metastatic potential correlates with enzymatic degradation of basement membrane collagen, *Nature (London),* 284, 67, 1980.
20. Jones, P. A. and DeClark, Y. A., Destruction of extracellular matrices containing glycoproteins, elastin and collagen by metastatic human tumor cells, *Cancer Res.*, 40, 3222, 1980.
21. Liotta, L. A., Lanzer, W. L., and Garbisa, S., Identification of a type V collagenolytic enzyme, *Biochem. Biophys. Res. Commun.*, 98, 184, 1981.
22. Felix, H. and Stauli, P., Different distribution of 100-A filaments in resting and locomotive leukemia cells, *Nature (London),* 261, 604, 1976.
23. Staruli, P. and Weiss, L., Cell locomotion and tumor penetration. Report on a workshop of the EORTC Cell Surface Project Group, *Eur. J. Cancer,* 5, 1, 1977.
24. Mareel, M. M., Malignant and nonmalignant cells: structural similarities and behavioural differences, *Experientia,* 36, 510, 1980.
25. Ludwig, J. and Titus, J. L., Experimental tumor cell emboli in lymph nodes, *Arch. Pathol.*, 84, 304, 1967.
26. Zeidman, I. and Buss, J. M., Experimental studies on the spread of cancer in the lymphatic system, *Cancer Res.*, 14, 403, 1954.
27. Hewitt, H. B. and Blake, E., Quantitative studies of translymphoidal passage of tumour cells naturally disseminated from a non-immunogenic murine squamous carcinoma, *Br. J. Cancer,* 31, 25, 1975.
28. Madden, R. E. and Gyune, L., Translymphoidal passage of tumor cells, *Oncology,* 22, 281, 1968.
29. Fisher, B. and Fisher, E. R., Barrier function of lymph node to tumor cells and erythrocytes, *Cancer,* 20, 1907, 1967.
30. Hanna, N. and Fidler, I. J., Role of natural killer cells in the destruction of circulating tumor emboli, *J. Natl. Cancer Inst.*, 65, 801, 1980.
31. Davey, G. C., Currie, G. A., and Alexander, P., Immunity as the predominant factor determining metastasis by murine lymphomas, *Br. J. Cancer,* 40, 590, 1979.
32. Mantovani, A., Jerrells, T. R., Dean, J. H., and Herberman, R. B., Cytolytic and cytostatic activity on tumor-cells of circulating human monocytes, *Int. J. Cancer,* 23, 18, 1979.
33. Fidler, I. J., The relationship of embolic homogeneity, number, size and viability to the incidence of experimental metastasis, *Eur. J. Cancer,* 9, 223, 1973.
34. Thompson, S. C., The colony forming efficiency of single cells and cell aggregates from a spontaneous mouse mammary tumour using the lung colony assay, *Br. J. Cancer,* 30, 332, 1974.
35. Kojima, K. and Sakai, I., On the role of stickiness of tumor cells in the formation of metastases, *Cancer Res.*, 24, 1887, 1964.
36. Winkelhake, J. L. and Nicolson, G. L., Determination of adhesive properties of variant metastatic melanoma cells to BALB/3T3 cells and their virus-transformed derivatives by a monolayer attachment assay, *J. Natl. Cancer Inst.*, 56, 285, 1976.
37. Zeidman, I., The fate of circulating tumor cells. I. Passage of cells through capillaries, *Cancer Res.*, 21, 38, 1961.

38. Sato, H., Khato, J., Sato, T., and Suzuki, M., Deformability and filtrability of tumor cells through "nucleopore" filter, with reference to viability and metastatic spread, in *Cancer Metastasis, Approaches to the Mechanism, Prevention and Treatment,* Stansly, P. G. and Sato, H., Eds., University of Tokyo Press, Tokyo, 1977, 53.
39. Buck, R. C., Walker 256 tumor implantation in normal and injured peritoneum studied by electron microscopy, scanning electron microscopy, and autoradiography, *Cancer Res.,* 33, 3181, 1973.
40. Warren, B. A. and Vales, O., The adhesion of thromboblastic tumour emboli to vessel walls *in vivo, Br. J. Exp. Pathol.,* 53, 301, 1972.
41. Ausprunk, D. H., Tumor angiogenesis, in *Chemical Messengers of the Inflammatory Process,* Houck, J. C., Ed., Elsevier/North Holland, New York, 1979, 317.
42. Schabel, F. M., Jr., Concepts for the systemic treatment of micrometastases, *Cancer,* 35, 15, 1975.
43. Folkman, J., Merler, E., Abernathy, C., and Williams, G., Isolation of a tumor factor responsible for angiogenesis, *J. Exp. Med.,* 133, 275, 1971.
44. Folkman, J., Tumor angiogenesis, *Adv. Cancer Res.,* 19, 331, 1974.
45. Folkman, J., Langer, R., Linhardt, R. J., Handenschild, C., and Taylor, S., Angiogenesis inhibition and tumor regression caused by heparin or a heparin fragment in the presence of cortisone, *Science,* 221, 719, 1983.
46. Dexter, D. L., Lee, E. S., DeFusco, D. J., Libbey, N. P., Spremulli, E. N., and Calabresi, P., Selection of metastatic variants from heterogeneous tumor cell lines using the chicken chorioallantoic membrane and nude mouse, *Cancer Res.,* 43, 1733, 1983.
47. Calabresi, P., Cancer, crab or chimera? The clinical implications of cancer cell heterogeneity, *Trans. Am. Clin. Climatol. Assoc.,* 92, 49, 1980.
48. Fidler, I. J., Metastasis: quantitative analysis of distribution and fate of tumor emboli labeled with ^{125}I-5-iodo-2'deoxyuridinine, *J. Natl. Cancer Inst.,* 45, 775, 1970.
49. Fidler, I. J., The relationship of embolic homogeneity, number, size and viability to the incidence of experimental metastasis, *Eur. J. Cancer,* 9, 223, 1973.
50. Coats, J., Metastasis of tumours (reviewer's comment), *Glasgow Med. J.,* 10, 330, 1878.
51. Paget, S., The distribution of secondary growths in cancer of the breast, *Lancet,* 1, 571, 1889.
52. Budd, W., The pathology and causes of cancer, *Lancet,* 2, 266, 1842.
53. Dickinson, W. H., Subcutaneous cancer, secondary to cancer of the glans penis, *Trans. Pathol. Soc. London,* 14, 242, 1863.
54. Viadana, E., Bross, I. D. J., and Pickren, J. W., An autopsy study of some routes of dissemination of cancer of the breast, *Br. J. Cancer,* 27, 336, 1973.
55. Sugarbaker, P. H., MacDonald, J. S., and Gunderson, L. L., Colorectal cancer, in *Cancer: Principles and Practices of Oncology,* DeVita, V., Jr., Hellman, S., and Rosenberg, S. A., Eds., Lippincott, Philadelphia, 1982, 643.
56. Moertel, C. G., The stomach, in *Cancer Medicine,* Holland, J. F. and Frei, E., III, Eds., Lea & Febiger, Philadelphia, 1973, 1527.
57. Prout, G. R., Jr., Prostate gland, in *Cancer Medicine,* Holland, J. F. and Frei, E., III, Eds., Lea & Febiger, Philadelphia, 1973, 1680.
58. Klein, E., Gradual transformation of solid into ascites tumors. Permanent difference between the original and the transformed sublines, *Cancer Res.,* 14, 482, 1954.
59. Klein, E., Gradual transformation of solid ascites tumors: evidence favoring the mutation selection theory, *Exp. Cell Res.,* 8, 188, 1955.
60. Fidler, I. J. and Kripke, M. L., Metastasis resulting from preexistant variant cells within a malignant tumor, *Science,* 197, 893, 1977.
61. Luria, S. E. and Delbruck, M., Mutations of bacteria from virus sensitivity to virus resistance, *Genetics,* 28, 491, 1943.
62. Kripke, M. L., Gruys, E., and Fidler, I. J., Metastatic heterogeneity of cells from an ultraviolet light-induced murine fibrosarcoma of recent origin, *Cancer Res.,* 38, 2962, 1978.
63. Fidler, I. J. and Hart, I. R., The origin of metastatic heterogeneity in tumors, *Eur. J. Cancer,* 17, 487, 1981.
64. Schirrmacher, V., Dzarlieva, R., Altevogt, P., Fogel, M., Waller, C. A., Dennis, J. W., Springer, G. F., Vlodavsky, I., Kramer, M., and Cheingsong-Popou, R., Phenotypic and genotypic differences between high- and low-metastatic related tumor lines and the problem of tumor progression and variant generation, in *Cancer Invasion and Metastasis: Biologic and Therapeutic Aspects,* Nicolson, G. L. and Luka, M., Eds., Raven Press, New York, 1984, 81.
65. Nicolson, G. L., Irimura, T., Nakajima, M., and Estrada, J., Metastatic cell attachment to and invasion of vascular endothelium and its underlying basal lamina using endothelial cell monolayers, in *Cancer Invasion and Metastasis: Biologic and Therapeutic Aspects,* Nicolson, G. L. and Luka, M., Eds., Raven Press, New York, 1984, 145.
66. Starkey, J. R., Hosick, H. L., Stanford, D. R., and Liggitt, H. D., Interaction of metastatic tumor cells with bovine lens capsule basement membrane, *Cancer Res.,* 44, 1585, 1984.

67. Carlsen, S. A., Ramshaw, I. A., and Warrington, R. D., Involvement of plasminogen activator production with tumor metastasis in a rat model, *Cancer Res.*, 44, 3012, 1984.
68. Talmadge, J. E., Wolman, S. R., and Fidler, I. J., Evidence for the clonal origin of spontaneous metastases, *Science,* 217, 361, 1982.
69. Fidler, I. J. and Hart, I. R., Biological and experimental consequences of the zonal composition of solid tumors, *Cancer Res.*, 41, 3266, 1981.
70. Reeve, J. G and Twentyman, P. R., Ploidy distribution of tumour cells derived from induced and spontaneously arising metastases of murine radiation-induced sarcoma, RIF-1, *Eur. J. Cancer Clin. Oncol.*, 18, 1001, 1982.
71. Spremulli, E. N., Dexter, D. L., Young, P., Campbell, D., and Calabresi, P., Clonal origin of spontaneous metastases of a human colon carcinoma xenografted in nude mice, *Proc. Am. Assoc. Cancer Res.*, 24, 28, 1983.
72. Poste, G., Doll, J., Brown, A. E., Tzeng, J., and Zeidman, I., Comparison of the metastatic properties of B16 melanoma clones isolated from cultured cell lines, subcutaneous tumors, and individual lung metastases, *Cancer Res.*, 42, 2770, 1982.

Chapter 5

QUANTITATIVE ASPECTS OF TUMOR HETEROGENEITY

"The goal... is to address this phenomenon of heterogeneity as it relates to understanding both ...*in situ* tumor growth dynamics *per se* and the tumor kinetic response to therapeutic agents."[1]

I. INTRODUCTION

In earlier chapters in this book (particularly Chapters 1 and 2), we have described the overall phenomenon of tumor heterogeneity. The complexity of this situation is staggering, and as a consequence, sophisticated insights should be coupled with quantitative methodologies to attempt to characterize and understand such complexity.

Therefore, this chapter will describe various analytical approaches that are being used in the quantitation of tumor heterogeneity. It should be appreciated by the reader that this is a young field with respect to the specific application of analytical methodology to the study of intraneoplastic diversity. Still, such methodologies will be crucial in the understanding, and hopefully, manipulation of this complex phenomenon.

At the onset, it is critical that the reader appreciate that the term "tumor heterogeneity" engenders different ideas in different groups of scientists. While this book has been primarily devoted to *cellular heterogeneity*, specifically the clonal presence and evolution of tumor subpopulations, there must also be understanding that *cell kinetic heterogeneity* is also present and is of great significance. Such cell kinetic heterogeneity would then explicitly mean cell cycle related compartments, within which all subpopulations reside. Clearly then, the time-related aspects of heterogeneity are major considerations which will play a role in the therapeutic application of cytotoxic modalities.

At this point, it is valuable to introduce an interactive representation of tumor heterogeneity as illustrated in Figure 1. In this Figure the major aspects of heterogeneity are schematically shown. This chapter will discuss aspects of heterogeneity associated with cell kinetics, but it should not be forgotten that this level of diversity overlaps and interacts with the other levels (e.g., Chapter 6).

II. ANALYTICAL ASPECTS OF TUMOR HETEROGENEITY

An attractive feature of research on intraneoplastic diversity is the implication that it is amenable to quantitative description, and emphasizes the view of tumor heterogeneity as a dynamic process. This dynamism is an inherent feature of the concept of tumor progression,[2,3] but encompasses much more than just the emergence of variant subpopulations as a function of growth. Clearly, such factors as the local microenvironment (e.g., pH, oxygen concentration, etc.) will influence the overall state of the tumor. There are both classical and new methodologies which are being applied to these problems, including flow cytometry,[1,4-18] cell separation by centrifugal elutriation, and other methods.[19-24] These types of analytical approaches requires some description prior to the discussion of their application.

A. Dissociation and Cell Separation Procedures

A basic problem in the biology of solid tumors is that they are composed of numerous subpopulations, i.e., intact malignant cells (which may of course be heterogeneous themselves in terms of differential phenotypic expression), various degenerating cells (i.e., from necrotic regions), cellular debris, and various normal tissue diploid cells

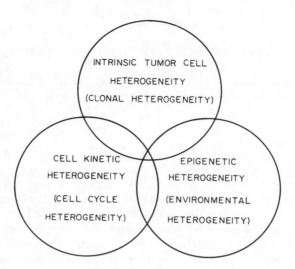

FIGURE 1. Schematic diagram illustrating the interacting compartments comprising overall tumor heterogeneity.

(e.g., macrophages, lymphocytes, fibroblasts, and endothelial and epithelial cells). To quantitate such cellular diversity, it is often necessary to dissociate the solid tumor into its component parts and, in some cases, to perform selective enrichment procedures relevant to the study of specific cellular subpopulations. Generally, the basic procedure is to excise the neoplasm in a sterile manner, mince it into fine pieces using sharp scissors or opposed scalpel blades, and enzymatically treat the minced pieces at 37°C for some period of time to obtain a truly single cell population.

This initial dissociation step is a relatively critical process, and many investigators have spent much effort in determining optimum enzyme(s) and incubation times so that the largest yield (i.e., cells per mg of tumor tissue) with the best viability is obtained.[19-24] As the cell subpopulations within a solid tumor can vary significantly in size, different zones may dissociate into single cells at different rates and with different efficiencies.[25] This is illustrated in Figure 2, where the cumulative yields (cells per mg) of dissociated solid carcinomas grown as xenografts in nude mice are shown. In this figure, the clones A and D represent specific subpopulations isolated from a heterogeneous human colon carcinoma (DLD-1).[26-28] These subpopulations (A and D) have been shown to constitute the major populations of the parental DLD-1 adenocarcinoma[26] (see Chapter 2). When two different types of enzymatic treatment are compared (i.e., a "cocktail" of DNAse I, collagenase, and pronase vs. trypsin), there are differences in both the *rate* of dissociation and in the total numbers of cells finally obtained. If such variables were not investigated, it would be possible to select a sampling time period unsuitable for the purposes of investigating neoplastic heterogeneity. For example, treating heterogeneous DLD-1 tumors for 10 min with 0.5% trypsin (Figure 2B) would preferentially select for the clone D subpopulation. As many investigators have pointed out that it is *vital* to obtain as near to a truly representative population of cells as is possible,[19-25] this initial disaggregation step represents a critical point as subsequent studies (flow cytometry, etc.) and the data interpretations therefrom will depend on this initial selection.

Cell yield after dissociation is only one of a number of criteria that is important in studies of the biology of heterogeneity. A second important aspect involves the viability and clonogenicity of cells. This can be illustrated in Figure 3. It can be seen that the two enzymatic treatments produce different numbers of clonogenic cells as a function

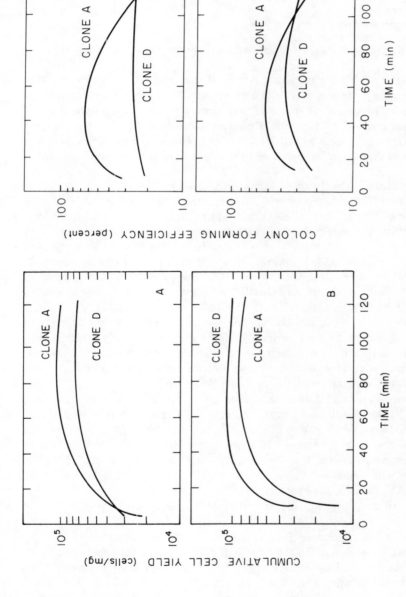

FIGURE 2. Cumulative cell yield (cells per mg) as a function of time for dissociation of clone A or clone D human colon adenocarcinomas. In A, the data for tumor treated with an enzyme cocktail containing 0.02% DNAse I, 0.02% collagenase, and 0.5% pronase are shown for the two solid tumors. In B, the data are for tumors treated with 0.5% trypsin.

FIGURE 3. Colony forming efficiency as a function time for dissociation of clone A or clone D human colon carcinomas. In A, data for tumors treated with an enzyme cocktail containing 0.02% DNAse I, 0.02% collagenase, and 0.5% pronase are shown for the two solid tumors. In B, the data are for tumors treated with 0.5% trypsin.

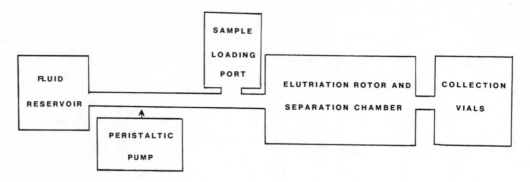

FIGURE 4. Schematic illustration of the system used in the centrifugal elutriation for cell separation.

of time, and have different kinetic patterns. Again, if such data is not known, it may lead to potential misinterpretation of subsequent results. This cannot be stressed too strongly, as an important aspect of intrinsic heterogeneity lies in the application of therapeutic (i.e., cytotoxic) modalities (c.f., Chapters 8 and 9). If baseline (i.e., unperturbed) values for parameters such as cell yield or clonogenicity are not investigated, it is unlikely that unambiguous data on perturbed heterogeneous tumors can be obtained.

B. Centrifugal Elutriation in the Study of Heterogeneous Systems

In investigations of cellular heterogeneity, there are a large number of biological problems that can be addressed, both theoretical and practical. Some of these problems can be more easily studied if it is possible to isolate (relatively) homogeneous cell subpopulations from, for example, an initial dissociate from a heterogeneous neoplasm. Centrifugal elutriation represents a technique by which such population homogeneity can be obtained, and has been described since 1948.[29] However, it has only recently become generally available with the introduction of appropriate technology, specifically the introduction of the "elutriator" rotor by the Beckman company.[30]

The basic principle of cell separation using this procedure is based on a balance of forces between an outwardly directed centrifugal force and an inwardly directed flow of fluid. As cellular sedimentation velocity is strongly influenced by cell diameter, centrifugal elutriation provides a basis for separation of cells based on size. A basic diagram of a centrifugal elutriation system is shown in Figure 4.

A single cell suspension is loaded into the separation chamber in the elutriation rotor at a combination of rotor speed and counter-current flow rate such that all cells are held in the chamber. Then, according to the needs of the investigation, either the rotor speed can be diminished (while keeping flow rate constant) or the flow rate can be increased (while keeping the rotor speed constant). As a consequence, the change in the balance of forces will then elute the smallest and least dense cell initially. By careful stepwise changes in the flow rate (or rotor speed), separation of cell populations from smallest to largest can be achieved, and selection of a homogeneous population (in size) is possible.[30,31]

The separation of a heterogeneous population is useful in many respects. For example, in Figure 5 we have also indicated how a fraction of radioactively labeled cells (with ^{3}HTdR) may be defined by autoradiography of the separated fractions. This illustrates that one could identify a markedly enriched fraction consisting of mainly S phase cells, and points out a major use of centrifugal elutriation studies — that of research concerned with cell cycle effects.[31,32] By combining centrifugal elutriation with other methodologies such as flow cytometry (*vide infra*), one may attempt to achieve a complete description of cell cycle-related heterogeneity. As tumor dissociations and

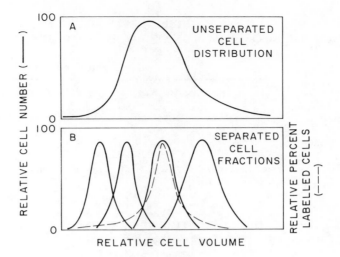

FIGURE 5. Illustration of the use of the centrifugal elutriation technique to obtain fractions of cells differing in cell volume. Also indicated in the Figure is the distribution of radioactively labeled cells (i.e., S phase cells) for each fraction, illustrating how centrifugal elutriation can be used for separation of cells throughout the mammalian cell cycle.

subsequent centrifugal elutriation separations can be done under sterile conditions, every fraction isolated can be used for further studies (i.e., cytotoxicity through the cell cycle).

A different use of centrifugal elutriation is concerned with the elucidation of the proportions of neoplastic vs. normal tissue cells, as these have been shown to vary among different tumor types.[33-38] As cells such as macrophages have been a recent focus of strategies of tumor eradication,[39-47] this is an important use of centrifugal elutriation. For example, Siemann et al.[48] studied the extent of host cell infiltration into three different rodent neoplasms (EMT6/R_o, KHT, and 9L/R_o) using centrifugal elutriation. Cell types isolated from eluted, dissociated solid tumors were identified from fractions by staining and microscopical examination, and the percentages found are listed in Table 1. It is apparent that the percentage of neoplasms that are actually parenchymal tumor cells varies from 37 to 60%, i.e., these rodent neoplasms contain a very large component of normal tissue cells. Given such diversity, it would not be surprising to expect that the intratumor distribution of normal tissue elements (e.g., macrophages) could also vary, particularly if tumors are best characterized by the "zonal" concept advanced by Fidler and Hart.[49] Siemann et al.[48] also demonstrated that the type of enzymatic treatment used for disaggregation (e.g., trypsin vs. an enzyme cocktail of DNAse, collagenase and pronase, and pronase vs. protease IX) had little effect on the ratio of host/tumor cells. However, differences in cell yield were noted, a finding also shown in the EMT6 tumor system by Twentyman and Yuhas.[50] Centrifugal elutriation was used to specifically separate host cells from cancer cells, and it was found that this easily could be achieved. The host cells due to their small size and generally less dense nature (e.g., the presence of highly vacuolated cytoplasm in macrophages) were removed preferentially in the early elutriation fractions. Such studies are attractive, as they suggest quantitative approaches to the study of interactions between normal and neoplastic cells in vivo (Chapter 6) and provide a way to kinetically follow dynamic shifts in composition during growth or after cytotoxic treatment.

Table 1
CELLULAR COMPOSITION OF RODENT SOLID TUMORS[a]

Tumor type	% Cellular composition			
	Tumor cells	Macrophages	Lymphocytes	Granulocytes
9L (rat)	54	19	19	8
EMT6 (mouse)	37	38	12	13
KHT (mouse)	60	31	6	3

[a] Data taken from Siemann, D. W., Lord, E. M., Keng, P. C., and Wheeler, K. T., *Br. J. Cancer,* 44, 100, 1981.

Table 2
CLONOGENICITY OF SUBPOPULATIONS OF CELLS SEPARATED FROM A MOUSE FIBROSARCOMA IN A RENOGRAFIN GRADIENT[a]

Population	Absolute lung tumor colony forming efficiency (%)
Unseparated tumor cells	1.53
Band 1 (density = 1.06)	3.95
Band 2 (density = 1.08)	5.35
Band 3 (density = 1.11)	2.94
Band 4 (density = 1.14)	2.88
Band 5 (density = 1.17)	1.07

[a] Data taken from Grdina, D. J., Milas, L., Mason, K. A., and Withers, H. R., *J. Natl. Cancer Inst.,* 52, 253, 1974.

C. Other Cell Separation Procedures

While centrifugal elutriation, as discussed above, is a methodology for the fast separation of large numbers of sterile, viable cells, there exist a wide variety of other cell separation techniques, some of which we will illustrate in this section.

For example, Grdina et al.[51-56] and Brock et al.[57] have described a procedure for the separation of at least five cellular subpopulations from a murine fibrosarcoma by equilibrium density centrifugation using Renograffin as the separation medium. This technique essentially separates cells on the basis of density. It has been shown that these subpopulations differ in clonogenic ability as shown by a lung colony assay,[53] and in their responses to ionizing radiation of different qualities (i.e., X-rays or neutrons).[51,54,55] Table 2 lists the relative clonogenic efficiencies of these separated subpopulations. These authors have also shown that the denser cell subpopulations may illustrate another important aspect of tumor cell heterogeneity, that being the presence of environmentally altered (hypoxic) cells.[55,56] As the presence of such hypoxic subpopulations has been proposed to create an environment in which certain therapeutic modalities (i.e., ionizing radiation) are ineffective,[57] this is an important methodological observation.

Durand[58] has performed experiments in which cells have been isolated from multicellular spheroids using tryspin dissociation and then separated into fractions using

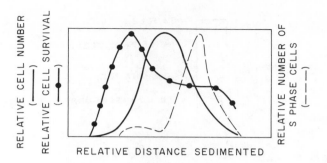

FIGURE 6. Illustration of the use of sedimentation of cells at unit gravity to achieve separation of cell subpopulations. In the figure, the distribution of cells is indicated together with the relative number of S phase cells in the distribution. The curve for the response of cells to ionizing radiation indicates the position of hypoxic cells.

differential sedimentation at unit gravity as the basis of the separation.[59] This technique will separate cells primarily on the basis of size, although density makes a minor contribution.[59] Data from his experiment are shown in Figure 6. These were interesting experiments, as they interrelated several factors. First, the data were obtained from multicellular spheroids, which are an in vitro model of solid tumors.[58,60] As can be seen in the figure, uptake of ^{3}HTdR was confined to the larger cells, with smaller cells not taking up radioisotope. This finding relates to the structure of a multicellular spheroid or nodular carcinoma in which there is a rim of proliferating cells surrounding an inner region of either non- or slowly proliferating cells plus possibly a necrotic zone. When cell subpopulations were exposed to a high dose of ionizing radiation (i.e., 22 Gy), it was found that the highest survival was for the smaller cells. This would indicate that these small cells, which were likely in G_1, were also hypoxic and therefore radiation resistant. This study illustrates the complexity of kinetic factors that must be considered in studies of intratumor heterogeneity.

D. Flow Cytometric (FCM) Techniques

The use of flow cytometric techniques in the analysis and interpretation of data has become invaluable and is being used in increasingly more sophisticated ways. The instrumentation is conceptually straightforward, as shown in Figure 7, and in its most basic description consists of a stream of single particles which are exposed to an optical signal. The type of optical signal can vary (e.g., visible light or laser generated light of varying frequency), but when it impinges on the target (i.e., the single cell stream) prior to analysis with a fluorescent dye, this dye will be excited when struck by a laser beam of appropriate frequency and will then fluoresce. This fluorescence can be monitored with appropriate optical sensors, and the intensity of the signal quantitated by electronic amplification and computation. If, for example, the amount of dye inside the cell is stoichiometrically related to the concentration of a particular molecule, it then would be possible to achieve a quantitative description of the frequency of occurrence and relative distribution of that molecule within the heterogeneous population under investigation. A wide variety of endpoints are available using such instrumentation, and a listing of potential applications which can relate to the use of flow cytometry is shown in Table 3, from which it is evident that there are many applications which can relate to the use of flow cytometry in tumor biology and the study of aspects of heterogeneity.

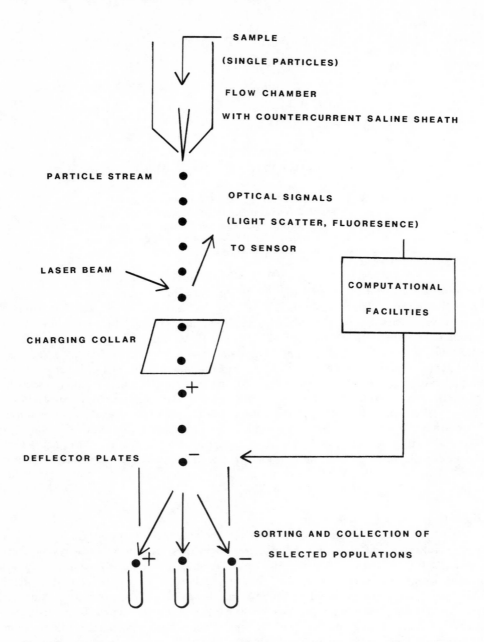

FIGURE 7. Illustration of the instrumentation used in the analysis of cell populations by flow cytometry. In the figure, the single cell suspension is illuminated by laser light, and the optical signal produced in the cells is recorded by the appropriate electronic instrumentation. As cells can be electrically charged according to whether a desired biological property (as denoted by the optical signal) is present, cells can be separated ("sorted") and collected for further analysis.

The simplest application of FCM is in the definition of the DNA content of a population of cells. As the mammalian cell cycle can be divided into four phases (specifically G_1, S, G_2, and M),[61] and as cells double their DNA content during the S phase, it is possible to discriminate between G_1 vs. S vs. G_2 or M cells based on their differences

Table 3
SOME USES OF FLOW CYTOMETRY IN TUMOR BIOLOGY

Physical parameter	Assayable biological property
Intensity of fluorescence	Cellular DNA content
	Cellular total protein content
	Activity of intracellular enzymes
	Assay of cellular viability
	Cellular uptake of certain drugs (e.g., adriamycin)
	Cellular permeability
	Phagocytosis
	Membrane potential
	Binding of antigens to cell receptors
	Analysis of chromatin structure
	Description of "Go" cells
Low angle light scatter Intensity (2—12°)	Cell viability
	Aspects of cell "size"
	Antibody capping
	Cell volume
90° angle light scatter	Elements of internal cellular architecture
Fluorescence decay time	Internal cellular microenvironment
Polarized fluorescence	Membrane fluidity

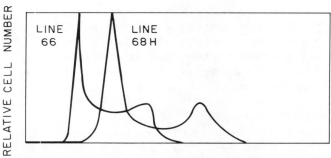

FIGURE 8. A plot of the relative DNA contents of two tumor subpopulations from a heterogeneous mouse mammary adenocarcinoma. (From Dethlefsen, L. A., Bauer, K. D., and Riley, R. M., *Cytometry*, 1, 89, 1980.)

in DNA content. Therefore, by looking only at a single parameter (i.e., DNA content per cell) and using FCM, large numbers of cells can be analyzed and the DNA distribution obtained.

In Figure 8, data on the DNA content of two subpopulations of tumor cells from the heterogeneous mouse mammary adenocarcinoma system of Heppner and co-workers is shown[1,62,63] (see Chapters 7 and 8 for additional discussion of this model system). These two subpopulations differ significantly in their DNA content with line 68H possessing roughly about 1.6 times the DNA of line 66 (based on the ratios of the G_1 DNA peaks). Heppner has reported that the modal chromosome number in line 68H is 130,

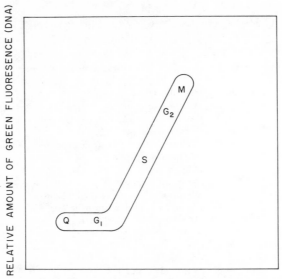

FIGURE 9. A scattergram of the distribution of cells possessing specific quantities of DNA and RNA using the acridine orange stain. This approach is used to identify specific cell cycle compartments.

while the modal number in line 66 is 39.[64] A similar finding has been noted for two subpopulations of cells isolated from a human colon carcinoma (DLD-1), termed clones A and D. These two subpopulations have been shown to have modal chromosome numbers of 76 (clone A) and 46 (clone D), and again, the ratio of the G_1 DNA contents of the two lines is about 1.6. These sets of data from two heterogeneous tumor model systems make it clear that differential DNA content can be used to identify and describe subpopulations. However, this by itself is not totally sufficient, as there are likely to exist subpopulations with similar DNA content, but which are dissimilar when other cellular properties are examined.

A technique in which two parameters (DNA and RNA content) are simultaneously determined for each cell has been developed using the stain acridine orange.[4,65,66] When cells are stained with the dye, and fluorescence is monitored at two different wavelengths (e.g., "red" or RNA fluorescence and "green" or DNA fluorescence), a plot can be made of cells with specific combinations of RNA and DNA as illustrated in Figure 9. As cells progress through the cell cycle, there is an increasing amount of RNA per cell. An important point is illustrated in the figure, in that quiescent ("Q") cells can be identified on the basis that although they have the same DNA content as their actively cycling counterparts, they have a *lower* RNA content. Indeed, the simplest kinetic definition of a heterogeneous neoplasm is one in which both P (proliferating) and Q (quiescent) cells are present.[1] The flow of cells from the Q to the P compartment (termed recruitment), particularly as it occurs after any type of cytotoxic treatment, is one of the most important considerations in the responsivity and curability of (heterogeneous) tumors. Further, the investigation of the presence and dynamics of the balance between Q and P cells may represent one of the most useful areas of the potential "clonal interactions" between tumor subpopulations (Chapter 7).

Flow cytometry has been used by Bauer et al.[67] to investigate whether such Q cells exist in the multicellular tumor spheroid model.[60] In this work, the EMT6 mouse sarcoma was used, and spheroids were enzymatically dissociated into single and the two-parameter DNA and RNA scattergram determined. The spheroids contained approximately 20% of cells that would have a DNA/RNA content consistent with being Q cells. Also, the smaller cells in the spheroid were the most likely to be Q-type cells, although Q cells were seen even among the largest cells. Further, these smaller Q cells tended to have a lower clonogenicity than did cells in other regions of the cell cycle. These data illustrate the interrelationships in regard to the cell cycle (i.e., is the tumor cell actively cycling, or is it a noncycling or Q cell), cellular DNA and RNA content, cell size, and potential ability to proliferate, and demonstrates the heterogeneity that exists in terms of cell kinetic considerations.

Wallen et al.[68,69] have recently investigated the growth kinetics of three of the mouse mammary adenocarcinoma subpopulations developed by Heppner and colleagues[62,63] (lines 66, 67, and 68H) using flow cytometry. The biology of these cell lines when in a quiescent state induced by nutrient deprivation in plateau phase cultures in vitro was studied. The quiescent state was shown to consist of more than 97% of cells in G_1 with a decrease of over 50% in RNA content. Further, cell volume decreased by about 50% and clonogenicity decreased markedly. The decrease in clonogenicity was more marked in line 66 than in lines 67 and 68H (factors of 7, 1.4, and 2.7, respectively), suggesting that line 66 would be the tumor line most strongly selected against in an in vivo heterogeneous neoplasm. Wallen et al.[68,69] found that the half-time for loss of clonogenicity was 32, 96, and 34 hr, respectively, for lines 66, 67, and 68H indicating the heterogeneity among these subpopulations. Also, the duration of the Q to P transition varied among the 3 cell lines, and it was found that line 67 reentered active proliferation most quickly, while lines 66 and 67 were significantly longer in their recruitment rates. Therefore "the ability of Q cells to re-enter the P state after a prolonged time in quiescence may be cell-line dependent."[69] Both of the factors described above (clonogenicity while in the Q state, and the rate of recruitment back to the P state) are important factors in both basic and applied aspects of the biology of neoplastic heterogeneity, and further research in this area is strongly indicated.

Another application of FCM has been performed by Lessin et al.,[6] who investigated some of the properties of the high and low metastatic variants of B16-F10 and B16-F1 from the heterogeneous B16 mouse melanoma system. Tumor cells were treated with fluorescamine (which penetrates the cell and stains both internal and external cell membranes) and with ethidium bromide (a DNA stain). Using the sorting capabilities of FCM equipment (Figure 7), the authors studied both membrane and DNA parameters. Further, using the scattering of light as a third parameter, cellular geometry could be compared to membrane and DNA parameters. Metastatic potential appeared to be correlated with subtle variations in membrane structure, and in particular was correlated with the appearance of a more condensed nuclear chromatin pattern. Enhanced metastatic capacity was also positively correlated with increased clonogenicity.[6,70] The authors also stated that metastatic arrest was influenced both by membrane and nuclear factors, because in the more metastatic line, there was much variability in the ratio of nuclear to total cell geometry, which was interpreted as an uncoupling between nuclear-DNA organization and cell membrane organization.[6,71] This was thought to be due to changes in microtubule/microfilament organization which occur during the process of cell transformation.[72]

In a similar study, Rameakers et al.[73] have used FCM to distinguish mixed cell populations by comparing the DNA content to the intracellular content of intermediate filament proteins (a class of cytoskeletal protein). This is an interesting approach, as not only is it useful in distinguishing tumor subpopulations, but it can differentiate host cells from tumor cells.

As an example of a sophisticated approach to studies of cellular heterogeneity with FCM, Sklarew[74,75] has investigated a rat sarcoma system comprised of four subpopulations with different chromosomal content. Using automated imaging methods for simultaneous determinations of DNA content with identification of ^{3}HTdR labeled cells, the four subpopulations could be identified and quantitated.

Because heterogeneous neoplasms are composed of both euploid and aneuploid subpopulations, the identification of subpopulations and knowledge of their proliferative activities is an important area for understanding cellular interactions and their control during tumor growth.

E. Cell Cycle Considerations in Heterogeneous Tumors

As illustrated by the work of Wallen et al.[68,69] and Sklarew,[74,75] sophisticated methodologies are being developed for the study of neoplastic heterogeneity. Still, it must be appreciated by the reader that at the present time, there are very little data dealing specifically with cell cycle alterations in heterogeneous tumors, either in the unperturbed situation or after cytotoxic treatment. It is, however, possible to present data to provide a framework within which to discuss potential cell kinetic aspects of heterogeneous cancers.

For example, a relevant way to approach such a discussion is to consider the intratumor kinetics that occur as a function of size.[76-78] In this regard, Watson[78] has studied the cell proliferation kinetics of the EMT6 mouse mammary sarcoma as a function of volume (Figure 10).

From Figure 10, it is obvious that there is a dramatic increase in the *measured* doubling time of the tumor with increasing volume (panel A). This increase is accompanied by an increase in the median cell cycle times of the tumor cells (from 14.1 hr for a 1.5-mm^3 volume to 18.5 hr for a 175-mm^3 volume), as well as a decrease in the rate of cell production. A slight increase in the rate of cell loss was also noted. With knowledge of the tumor cell cycle times, and the fractions of cells incorporating ^{3}HTdR (S phase cells) termed the labeling index, the "potential" doubling time of the tumor may be calculated as a function of volume (i.e., the tumor doubling time that would occur without any cell loss). This is illustrated in panel A of Figure 10. In panel C, the so-called cell loss factor (ϕ) is shown as a function of volume. This factor is essentially the ratio of the rate of cell loss to that of cell production (K_L/K_p), and is markedly influenced by volume. Further, the growth fraction of the tumor (i.e., the percentage of cells actively proceeding through the cell cycle) decreases significantly as tumor volume increases.

Figure 10 graphically shows that there are a number of areas in which differential cell kinetic parameters might exist among cell subpopulations and which could modify tumor growth. For example, aneuploid subpopulations of solid tumors tend to show higher proliferative activity than do diploid subpopulations. This finding illustrates that cell cycle times could vary among subpopulations. Also, the cell loss factor would be expected to exhibit variability, as it is based on the ratio of cell production to cell loss, either of which could vary among tumor subpopulations. Denekamp has shown that carcinomas tend normally to have significantly higher cell loss factors than do sarcomas.[79] Further, as different mechanisms of cell loss exist (host immune responses, apoptosis, metastasis, exfoliation), it may be that measurement of the cell loss factor could provide a significant, quantitative measure of intratumor heterogeneity.

In this regard, Humphries and Isaacs[80] have studied the influence of androgens on the cell kinetics of Dunning R-3327-G rat prostatic adenocarcinoma. These authors point out that an individual prostatic tumor could be heterogeneously composed of three distinctly different tumor phenotypes: i.e., androgen dependent; androgen inde-

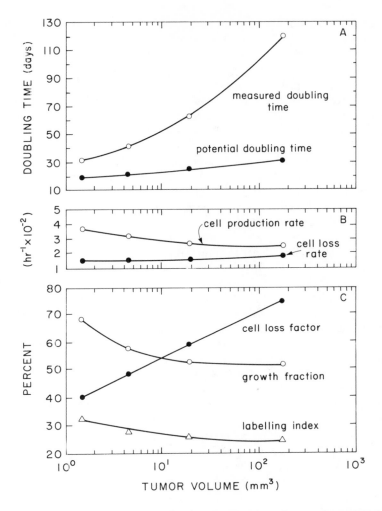

FIGURE 10. Changes in the cell kinetic parameters of a mouse mammary adenocarcinoma as the tumor increases in volume. (From Watson, J. V., *Cell Tissue Kinet.*, 9, 147, 1976,. With permission.)

pendent and insensitive; and androgen independent but sensitive,[80-82] in terms of whether androgens can stimulate growth. The authors found that the cell loss factor in control tumors that were androgen independent but sensitive was 0.36, and that this value increased to 0.74 in castrated rats. Cell death was therefore increased by a factor of 2 after castration, a situation that could be reversed by testosterone administration. In this tumor, the cells are highly sensitive to androgens as defined by cell loss. Therefore, cell kinetic mechanisms can be differentially affected in heterogeneous tumors by environmental factors (i.e., hormones).

III. CLINICAL EVIDENCE OF KINETIC TUMOR HETEROGENEITY

The following material will describe clinical tumor heterogeneity, mainly as evidence from FCM studies.[83,84] It is important to note that these data are concerned mainly with ploidy, and that essentially *no* data exist on the actual cell kinetics of heterogeneous cancers. These ploidy data are therefore unfortunately only static representations of dynamic neoplasms. An aim of the next generation of FCM and related investigations on intraneoplastic diversity should be to present a more detailed picture of the actual kinetic behavior of subpopulations within solid tumors.

A. Studies on Various Tumor Types

Barlogie et al.[5] have described the frequency and severity of abnormalities in ploidy in a large number of individuals having a variety of types of tumors, and have attempted to ascertain the prognostic significance of abnormal chromosome content (e.g., aneuploidy) and proliferative status. These studies included individuals with breast cancer, melanoma, gastrointestinal cancer, lung cancer, lymphoma, genitourinary cancer, and other miscellaneous tumor types. In some patients, it was also possible to study tumor ploidy in different *sites* (i.e., primary vs. metastases), over *time* (i.e., progression), both very important considerations in the evolution of tumor heterogeneity. While the percentage of aneuploid cells did vary among primary and metastatic sites, no significant differences in the extent of aneuploidy were seen in different disease sites within a given patient. In terms of the temporal stability of ploidy, in only one anaplastic carcinoma was a significant change in DNA content seen. Some of these data are presented in Figure 11. In Figure 11A, the stability of the average DNA contents of tumors in patients presenting at increasing time periods after diagnosis is shown, both for primary and metastatic cancers. There is a trend to higher DNA content in patients' tumors at about 1 year after diagnosis, with a decline out to greater than 4 years after diagnosis for the primary neoplasms. This could represent an aspect of tumor progression (i.e., more chromosomal abnormalities with time[85]). For the metastases examined, there is no trend in DNA content with time. In Figure 11B, the percentage of tumors that were aneuploid at the time of examination is shown. In this figure, there is also a trend to a greater degree of aneuploidy with increasing time (for both primary and metastatic disease), with a decline at longer observation times. The average DNA content of tumors from all patients was 1.4 times the DNA content of normal human granulocytes, and 81% of patients had tumor DNA contents clearly distinguishable from normal diploid cells.[85,86] Also 14 patients (7%) clearly had *at least* two discretely different aneuploid cell populations, and the more aneuploid tumors also had a higher percentage of cells in the S phase of the cell cycle. There appeared to be no relationship between the DNA content or proliferative activity of a primary neoplasm per se and appearance of metastatic disease although Cifone and Fidler[87] have described a positive correlation between increased tumor cell mutability (i.e., increased genomic instability) and the ability to form metastases. Greater aneuploidy was seen in patients with metastatic disease from lung, breast, and gastrointestinal cancer.

A study similar to that of Barlogie et al.[5] has been conducted by Johnson et al.[88] who studied the ploidy and DNA contents of spontaenous neoplasms in dogs. It was found that, on the average, the neoplasms contained a DNA content of about 1.4 times that of normal dog peripheral blood leucocytes. Melanomas, soft-tissue sarcomas, and squamous cell carcinomas had DNA contents close to the normal diploid values, whereas mammary carcinomas and chondro- and osteosarcomas tended to exhibit abnormal values of DNA content. Interestingly, 6 of 68 neoplasms exhibited the presence of more than two cellular populations, and these occurred only in mammary carcinomas and osteosarcomas. Of the six tumors showing the presence of more than two subpopulations, the DNA content of the second was about twice that of the first. Also, there was a tendency for a higher percentage of S phase cells to be found in poorly differentiated tumors with higher DNA contents. Therefore, these authors felt that there was a good correlation among the histological degrees of differentiation, invasiveness, extent of DNA ploidy elevation, and proportion of cells in the S phase. The

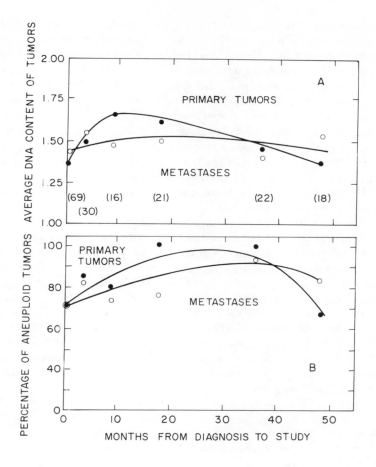

FIGURE 11. Changes in the average DNA content of human tumors (panel A), or in the percentage of aneuploid cells in the tumors (panel B), as a function of time after initial diagnosis. (From Barlogie, B., Johnston, D. A., Smallwood, L., Rober, M. N., Maddox, A. M., Latreille, J., Swartzendruber, D. E., and Drewinko, B., *Cancer Genet. Cytogenet.*, 6, 17, 1982. With permission.)

percentage of abnormal ploidy in the dog tumors (i.e., about 80%) compares well to values of 60 to 80% reported for human tumors.[89-94]

A large study has been done by Frankfurt et al.[95] on the DNA content of primary and metastatic human solid tumors. A total of 656 neoplasms (365 primary and 291 metastatic) were examined, and aneuploid cells were seen in 430 (66%). Thirty of these neoplasms had two aneuploid sublines and two had three aneuploid sublines. These were most frequently noted in colorectal carcinomas and in sarcomas. No major differences were seen in frequency or degree of aneuploidy between primary and metastatic cancers. In 15 instances, tissue from both the primary and metastatic lesions were obtained, and in only 3 of these were differences seen (20%). The changes noted occurred in a variable manner, as one patient exhibited a higher aneuploidy state in the metastasis, while a second showed a lower aneuploid state in the metastasis, while the third showed selection of one aneuploid subpopulation from a mixed diploid/aneuploid primary cancer.

Grassel and Horstein[17] have investigated the DNA contents of squamous carcinomas of the oral, pharyngeal, and laryngeal mucosa. In two patients, both the primary tumor and a metastasis to a lymph node could be studied. Most samples exhibited aneuploidy, with the average DNA content being 1.55 times the diploid content of normal tissue cells. In only one instance did the DNA histogram reveal a tumor with more than one aneuploid subpopulation. In two patients in which both primary and metastatic lesions were compared, a complete correspondence of DNA content was found.

Tumors of the bladder have been investigated by Jakobsen et al.[16,96] These cancers have typically been shown to contain aneuploid cell populations when they are of high-grade malignancy.[97] For higher grade tumors, about 97% were found to be aneuploid. Also, 15% were found to contain more than one aneuploid cell population. The authors concluded that a high degree of aneuploidy was, however, not necessarily associated with a high degree of invasiveness.

An interesting study has been performed by Friedlander et al.,[98] who took paraffin-embedded tumor tissues of ovarian cancer and performed flow cytometric studies which allowed access to material not previously available. Briefly, sections of embedded tissue were dewaxed, rehydrated, and digested with enzyme to obtain a single cell suspension. On FCM analysis, it was found that 31% of neoplasms contained only a single G_1 peak and were therefore classified as diploid. However, aneuploidy was found in 69% of the tumors, and 15% had more than one aneuploid population.

Raber et al.[11] have attempted to correlate the DNA content, proliferative activity, and estrogen receptor status in patients with breast cancer. Of tumors analyzed, 85% were considered aneuploid and appeared to consist of only two populations. In 9%, two distinct aneuploid subpopulations could be found. In five patients with metastatic disease who were repeatedly biopsied over a time period of 2 to 36 weeks, no changes were seen in DNA values. In these studies, estrogen receptor positive (ER^+) tumors were more frequently diploid than were ER^- cancers. Such neoplasms are also known to have a more favorable prognosis. It would appear, therefore, that the evolution of subpopulations with a higher degree of aneuploidy is also associated with a loss of estrogen receptor positivity.[99]

Tribukait et al.[100] have investigated human colorectal carcinomas using FCM. In 70% of patient, biopsies were taken from four quadrants of tumors, with a separation of at least 1 cm kept between specimens. A total of 124 tumor cell populations were found in 66 colorectal adenocarcinomas. Importantly, the authors state that as the majority of their biopsy specimens appeared to contain only one cell population, the different cell populations of multiploid tumors must occupy separate regions. This suggestion is consistent with the zonal composition concept.[49] The frequency of multiclonality was higher in poorly differentiated as compared to highly or moderately differentiated tumors.

Heliö et al.[101] have also shown marked aneuploidy in malignant bone tumors. About 70% of neoplasms were aneuploid, showing either one or two subpopulations. Further, about 50% of metastatic bone cancers were also aneuploid.

An important study recently has been done by Auer et al.[102] who investigated the progression of mammary adenocarcinomas using DNA content as the marker. The DNA contents of primary tumors were compared with those found in local and distant metastases up to 12 years later. In only one patient (1/18) was a change to a more aneuploid karyotype seen in the metastases, suggesting a high degree of stability of DNA content during the history of mammary adenocarcinoma.

B. Lung Tumors (Nonsmall Cell Carcinomas)

Nervi et al.[14] have made the statement that "the possibility of improving the therapeutic ratio in human tumors is strictly related to the knowledge of the biological

Table 4
VARIABILITY BETWEEN ANEUPLOID AND DIPLOID POPULATIONS IN BIOPSY SPECIMENS OF HUMAN LUNG TUMORS[a]

Tumor number	Number of samples assayed	Proportion of aneuploid to diploid cells (A/D, mean ± Standard Error)
1	4	0.57 ± 0.20
2	3	0.12 ± 0.02
3	3	0.11 ± 0.08
4	3	0.12 ± 0.04
5	3	0.08 ± 0.02

[a] Data taken from Nervi, C., Badaracco, G., Maisto, A., Mauro, F., Tirindelli-Danesi, D., and Starace, G., *Cytometry*, 4, 174, 1983.

structure of the tumor." In this regard, the authors have investigated the cytogenetic and proliferative heterogneity of human solid tumors, specifically nonsmall cell carcinoma (NSCLC) of the lung. It was found that with the exception of only *one* tumor (an epidermoid carcinoma metastatic to lung), all other tumors showed multiple cellular populations. Of the neoplasms exhibiting multiple cell populations, all of them also possessed a population with a normal, diploid DNA value which could represent either a tumor cell clone with a diploid DNA content, or simply normal tissue cells (i.e., macrophages) in the biopsy specimen.

In the work of Nervi et al.,[14] multiple tumor biopsy specimens were taken from different areas so that DNA content from area to area could be evaluated. This study also bears directly on the zonal composition concept of Fidler and Hart.[49] Nervi et al.[14] did indeed find significant variation (4 to 5-fold) in the relative proportions of diploid to aneuploid cells when different areas were compared (Table 4).

Other work on NSCLC tumors has been reported by Teodori et al.[15] As in the work of Nervi et al.,[14] the authors identified in epidermoid carcinomas, tumor subpopulations with both diploid and aneuploid DNA content. Also, both hypoploid and hyperdiploid clonal populations could be found. Of 23 patients, the total number of subpopulations noted in biopsy specimens was 59 (i.e., 2 to 3 subpopulations per tumor). The average value of the DNA content was about 1.34. In adenocarcinoma of the lung, all tumors showed multiple subpopulations.[15] In 14 tumors, the average number of subpopulations per tumor was about 3 (i.e., 43 subpopulations in 14 tumors), and the average value of the DNA content was about 1.67 with a range of 0.8 to 3.0. The adenocarcinomas on the average appeared to have a higher DNA content than other forms of NSCLC.

For patients with large-cell carcinoma of the lung, all tumors showed multiple subpopulations. In 12 cases, the average number of subpopulations per tumor was about 2 (i.e., 26 populations). The average value of the DNA content was 1.4.

Teodori et al.[15] have continued the work of Nervi et al.,[14] in terms of describing the internal variability of tumors. For example, in one patient with epidermoid carcinoma, an FCM sample from the core of tumor showed the presence of two subpopulations, while a sample taken from the periphery showed the presence of three subpopulations (the two present in the core sample and one additional, more aneuploid population). Additionally, when three different lymph node biopsy specimens were analyzed, one showed one neoplastic subpopulation, another showed two subpopulations, while a third also showed two subpopulations, but not the same populations as in the second lymph node. Therefore, not only is intratumoral sampling of concern (based on random sampling variations), but there may be significant diversity in extent of tumor progression as seen from lymph node analysis.

A second example is given by Teodori et al.[15] of differential sampling from an epidermoid carcinoma, in which four subpopulations were found in the tumor core (two hypodiploid, one diploid, and one hyperdiploid), whereas only three subpopulations were found at the tumor periphery (the diploid and two different hyperploid populations). Of these populations identified in the core and periphery of this carcinoma, only three of them were found in a biopsied lymph node. These data further illustrate intratumor heterogeneity as well as the diversity that can be encountered in different sites.

Teodori et al.[15] also give data indicating an inverse relationship between tumor doubling time and DNA content. Solid tumors consisting only of diploid cells had doubling times of about 8 months. However, in NSCLC where there might be an average intraneoplastic DNA content of about 1.3 (about two to three subpopulations per tumor), the average doubling time decreased to about 7 months. In adenocarcinoma of the lung, where there might be an average DNA content of about 1.6 to 1.7 (with possibly three subpopulations per tumor), the average doubling time was about 5.5 months. Therefore, the more progressed (aneuploid) the tumor, the shorter the measured doubling time.

C. Lung Tumors (Small Cell Carcinomas)

Vindeløv and co-workers[7,8] have investigated aspects of small-cell carcinoma of the lung using FCM procedures. Such tumors have been shown to possess great variability in chromosome content, doubling times, and cell cycle-related parameters.[103,104] In this work, metastases were examined from sites such as skin and lymph nodes, and multiple biopsy specimens were examined over a period of weeks to months. In 23 of 29 metastases, only 1 tumor population could be detected, while in the other 6, evidence of 2 subpopulations with differing DNA content was found. In contrast to the findings of Teodori et al.,[15] when multiple metastases were compared, no differences were found between different sites. Vindeløv et al.[7] felt that subpopulations constituting less than 30% of all cells may be easily overlooked by FCM, and that subpopulations comprising 5 to 10% of a tumor likely can not be identified, even in very favorable situations.

In this section dealing with clinical and experimental evidence for quantitative tumor heterogeneity, we have described a number of studies having as a unifying feature the application of analytical methodology (i.e., FCM, etc.) to intraneoplastic heterogeneity. Some of these data have been discussed previously (Chapter 2), but in a different context. Here, we have attempted to quantitatively present the evidence for intratumor heterogeneity to provide a basis for further experimentation. As stated earlier in this chapter, little data on cell kinetic aspects of tumor heterogeneity actually exist, with the exception of the FCM studies. The FCM work documents that heterogeneity exists, and that spatial (e.g., intratumor progression) considerations are vital in the overall understanding of heterogeneity. Further, as we have stressed the dynamic aspects of tumor heterogeneity, it should be apparent that the next stage in the documentation of dynamic heterogeneity will include description of cell kinetic parameters within heterogeneous tumors (i.e., growth fractions, cell loss factors, percentages of host cells, etc.). These will complement and extend this first generation of essentially FCM data, which were extremely useful to document the existence of tumor heterogeneity, particularly in human cancers.

IV. SUMMARY

The use of flow cytometric analysis is clearly an important tool in the description of the cellular status of a solid tumor both as a static descriptor and from the view point of describing dynamic changes with time. For detailed descriptions of the mathematical and statistical limitations of FCM, the reader is referred to the work of Baisch et al.,[105] Dean and Jett,[106] Fried,[107] Gray,[108] Kim and Perry,[109] Salzman et al.,[110] and Shackney.[111]

The description of the existence of multiple subpopulations in tumor specimens, their variability from region to region, and the question of whether the primary neoplasm and its metastases contain the same or different subpopulations merit serious consideration. Another highly relevant question of analysis which has significant impact on any description of tumor dynamics is the basic question of "how many subpopulations actually exist in heterogeneous cancers?" The answer based on considerations of all possible phenotypic properties (as discussed in Chapters 1 and 2) is likely "very many". However, based on the FCM data as discussed in this chapter, there do not appear to exist a very large number of identifiable subpopulations within tumors (at least as based solely on DNA and RNA measurements). However, as FCM data are usually based upon only one parameter (DNA), it must be remembered that even within a "diploid" tumor, subpopulations exist if a different trait (e.g., response to cytotoxic agents) is studied.

There are subtleties involved in the assessment of numbers of subpopulations by current available methodologies. For example, as pointed out by Schuette et al.[112] and Ritch et al.,[113] the ability to resolve closely spaced peaks in a DNA histogram includes such factors as the absolute positions of the peaks, the coefficient of variation (CV) of the individual peaks, the relative heights of peaks (i.e., the percent of cells under each peak), the relative apposition of peaks, and very importantly, the computational power available. As we have noted previously Vindeløv et al.[7] feel that subpopulations comprising 5 to 10% of a tumor probably cannot be identified.

A direct illustration of this problem of the power to resolve tumor subpopulations is the work performed by Shapiro et al.,[114] who studied heterogeneity of human brain tumors. Biopsy specimens were dissociated into single cells, and chromosome studies were performed 2 to 3 days later. Multiple populations of cells could be distinguished based on chromosomal content. The authors determined how many subpopulations existed within each tumor based on the occurrence of a minimal number of cells with identical karyotypes. From 8 gliomas, they identified a total of 117 subpopulations, and on the average, each tumor contained 14 to 15 subpopulations. The range was 3 to 21 subpopulations per tumor. These numbers are obviously greater than the number typically identified by FCM analysis, and it is apparent that karyotypic analysis has greater resolving power than FCM per se, as small changes in chromosome composition may occur without any large changes in DNA content. Also, the resolving power of the karyotype approach is greater based to a large extent on the absolute cutoff used to identify karyotypically unique cells that qualify as constituting a new subpopulation. Shapiro et al.[114] used a minimum of 5 karyotypically identifiable cells (out of an average of about 360 metaphase cells analyzed) as evidence of a true subpopulation. A more rigid requirement (e.g., 25 karyotypically similar cells per tumor subpopulation) would bring the karyotypic information in line with the FCM data. Such a 25-cell requirement would then be roughly equivalent to identification of a subpopulation of approximately 7% of a total population.[7] These studies indicate that the answer to "how many subpopulations" for a given phenotype is largely defined by the technique used.

The FCM data appear to indicate that the upper limit of the number of subpopulations for karyotypic diversity is at most four to five (including a diploid population if present). More usually, two or three subpopulations are found, even in the most advanced cancers. It is worth noting that simply because a subpopulation is present as a minor component in an FCM study does *not* mean that the proliferative capacity of that subpopulation is also low. It is quite possible to encounter a situation in which aneuploid populations comprise a small percentage of the total population, but have fractions of cells in S phase considerably higher than that seen in the majority diploid population.[113] This should have impact on therapeutic strategies of heterogeneous tumors that employ cell cycle-specific cytotoxic agents as the implication is that these agents might be effective, but possibly only against a minor subpopulation.[115,116]

The future will provide exciting new studies on cell kinetic aspects of inter- and intratumor heterogeneity.[1,117,118] These are important studies, and are vital to the advancement of cancer therapy. A central feature of such studies will be the identification of Q and P cells within tumors, as these studies may yield the most basic and ultimately important definition of tumor heterogeneity.[68,69,117]

REFERENCES

1. Dethlefsen, L. A., Bauer, K. D., and Riley, R. M., Analytical cytometric approaches to heterogeneous cell populations in solid tumors: a review, *Cytometry,* 1, 89, 1980.
2. Nowell, P. C., The clonal evolution of tumor cell populations, *Science,* (Wash. D.C.), 194, 23, 1976.
3. Nowell, P. C., Tumors as clonal proliferation, *Virchows Arch. B: (Cell Pathol.),* 29, 145, 1978.
4. Darzynkewicz, Z. and Andruff, M., Multiparameter flow cytometry, Part 1: Applications in analysis of the cell cycle, *Clin. Bull.,* 11, 47, 1981.
5. Barlogie, B., Johnston, D. A., Smallwood, L., Rober, M. N., Maddox, A. M., Latreille, J., Swartzendruber, D. E., and Drewinko, B., Prognostic implications of ploidy and proliferative activity in human solid tumors, *Cancer Genet. Cytogenet.,* 6, 17, 1982.
6. Lessin, S. R., Abraham, S. R., and Nicolini, C., Biophysical indentification and sorting of high metastatic variants from B16 melanoma, *Cytometry,* 2, 407, 1982.
7. Vindeløv, L. L., Hansen, H. H., Christensen, I. J., Spang-Thomsen, M., Hirsch, F. R., Hansen, M., and Nissen, N. I., Clonal heterogeneity of small-cell anaplastic carcinoma of the lung demonstrated by flow-cytometric DNA analysis, *Cancer Res.,* 40, 4295, 1980.
8. Vindeløv, L. L., Hansen, H. H., Gersel, A., Hirsch, F. R., and Nissen, N. I., Treatment of small-cell carcinoma of the lung monitored by sequential flow cytometric DNA analysis, *Cancer Res.,* 42, 2499, 1982.
9. Zucker, R. M., Whittington, K., and Price, B. J., Differentiation of HL-60 cells: cell volume and cell cycle changes, *Cytometry,* 3, 414, 1983.
10. Piwnicka, M., Darzynkewicz, A., and Melamed, M. R., RNA and DNA content of isolated cell nuclei measured multi-parameter flow cytometry, *Cytometry,* 3, 269, 1983.
11. Raber, M. N., Barlogie, B., Latreille, J., Bedrossian, C., Fritsche, H., and Blumenschein, G., Ploidy, proliferative activity and estrogen receptor content in human breast cancer, *Cytometry,* 3, 36, 1982.
12. Beck, H. P., Brammer, I., Zywietz, F., and Jung, H., The application of flow cytometry for the quantification of the response of experimental tumors to irradiation, *Cytometry,* 2, 44, 1981.
13. Kamarck, M. E., Barbosa, J. A., Kuhn, L., Messer, Peters, P. G., Shulman, L., and Ruddle, F. H., Somatic cell genetics and flow cytometry, *Cytometry,* 4, 99, 1983.
14. Nervi, C., Badaracco, G., Maisto, A., Mauro, F., Tirindelli-Danesi, D., and Starace, G., Cytometric evidence of cytogenetic and proliferative heterogeneity of human solid tumors, *Cytometry,* 2, 303, 1982.
15. Teodori, L., Tirindelli-Danesi, D., Mauro, F., DeVita, R., Uccelli, R., Botti, C., Modini, C., Nervi, C., and Stipa, S., Non-small-cell lung carcinoma: tumor characterization on the basis of flow cytometrically determined cellular heterogeneity, *Cytometry,* 4, 174, 1983.
16. Jakobsen, A., Mommsen, S., and Olsen, S., Characterization of ploidy level in bladder tumors and selected site specimens by flow cytometry, *Cytometry,* 4, 170, 1983.
17. Grassel, R. and Hornstein, O. P., Flow cytometric measurement of ploidy and proliferative activity of carcinomas of the oropharyngeal carcinoma, *Arch. Dermatol. Res.,* 273, 121, 1982.
18. Reeve, J. G. and Twentyman, P. R., Ploidy distribution of tumour cells derived from induced and spontaneously arising metastases of a murine radiation-induced sarcoma, RIF-1, *Eur. J. Cancer Clin. Oncol.,* 18, 1001, 1982.
19. Barendsen, G. W., Analysis of tumour responses by excision and *in vitro* assay of clonogenic capacity, *Br. J. Cancer,* 41 (Suppl. IV), 209, 1980.

20. McNally, N. J. and deRonde, J., Radiobiological studies of tumours *in situ* compared with cell survival, *Br. J. Cancer,* 41 (Suppl. IV), 259, 1980.
21. Rasey, J. S. and Nelson, N. J., Response of an *in vivo — in vitro* tumour to X-rays and cytotoxic drugs: effects of tumour disaggregation method on cell survival, *Br. J. Cancer,* 41 (Suppl. IV), 217, 1980.
22. Rasey, J. S. and Nelson, N. J., Effect of tumor disaggregation on results of *in vitro* cell survival assay after *in vivo* treatment of the EMT-6 tumor: X-rays, cyclophosphamide, and bleomycin, *In Vitro,* 16, 547, 1980.
23. Rasey, J. S. and Nelson, N. J., Discrepancies between patterns of potentially lethal damage repair in the RIF-1 system *in vitro* and *in vivo, Radiat. Res.,* 93, 157, 1983.
24. Rockwell, S., *In vivo — in vitro* tumour cell lines: characteristics and limitations as models for human cancer, *Br. J. Cancer,* 41 (Suppl. IV), 118, 1980.
25. Leith, J. T., Bliven, S. F., Lee, E. S., Glicksman, A. S., and Dexter, D. L., Intrinsic and extrinsic heterogeneity in the responses of parent and clonal human colon carcinoma xenografts of X-irradiation, *Cancer Res.,* 44, 3757, 1984.
26. Dexter, D. L., Spremulli, E. N., Fligiel, Z., Barbosa, J. A., Vogel, R., VanVoorhees, A., and Calabresi, P., Heterogeneity of cancer cells from a single human carcinoma, *Am. J. Med.,* 71, 949, 1981.
27. Dexter, D. L. and Calabresi, P., Intraneoplastic diversity, *Biochem. Biophys. Acta,* 695, 97, 1982.
28. Dexter, D. L., Neoplastic subpopulations in carcinomas, *Ann. Clin. Lab. Sci.,* 11, 98, 1981.
29. Lindahl, P. E., Principle of a counter-streaming centrifuge for the separation of particles of different sizes, *Nature (London),* 161, 648, 1948.
30. Keng, P.C., Li, C. K. N., and Wheeler, K. T., Characterization of the separation properties of the Beckman elutriator system, *Cell Biophys.,* 3, 41, 1981.
31. Keng, P. C., Li, C. K. N., and Wheeler, K. T., Synchronization of 9L rat brain tumor cells by centrifugal elutriation, *Cell Biophys.,* 2, 191, 1980.
32. Preiser, H., Walczak, I., Renick, J., and Rustum, Y. M., Separation of leukemic cells into proliferative and quiescent subpopulations by centrifugal elutriation, *Cancer Res.,* 37, 3876, 1977.
33. Evans, R., Macrophages in syngeneic animal tumors, *Transplantation,* 14, 468, 1972.
34. Russell, S. W., Doe, W. F., Hoskins, R. G., and Cochrane, D. G., Inflammatory cells in solid murine neoplasms, I. Tumor disaggregation and identification of constituent inflammatory cells, *Int. J. Cancer,* 18, 322, 1976.
35. Stewart, C. C. and Meetham, P. L., Cytocidal activity and proliferation ability of macrophages infiltrating the EMT6 tumor, *Int. J. Cancer,* 22, 152, 1978.
36. Lord, E. M. and Keng, P. C., Effects of radiation of *in situ* host cells separated from a murine tumor by centrifugal elutriation, *Radiat. Res.,* 83, 456, 1980.
37. Bugelski, P. J., Kirsch, R. L., and Poste, G., New histochemical method for measuring intratumoral macrophages and macrophage recruitment into experimental metastases, *Cancer Res.,* 43, 5493, 1983.
38. Gandour, D. M. and Walker, W. S., Isolation and characterization of cell cycle-enriched subpopulations of a murine macrophage cell line by centrifugal elutriation, *Exp. Cell Res.,* 143, 327, 11983.
39. Fidler, I. J., Barnes, Z., Fogler, W. E., Kirsh, R., Bugelski, P., and Poste, G., Involvement of macrophages in the eradication of established metastases following intravenous injection of liposomes containing macrophage activators, *Cancer Res.,* 42, 496, 1982.
40. Fidler, I. J. and Poste, G., Macrophage-mediated destruction of malignant tumor cells and new strategies for the therapy of metastatic disease. *Springer Semin. Immunopathol.,* 5, 161, 1982.
41. Hibbs, J., Discrimination between neoplastic and non-neoplastic cells *in vitro* by activated macrophages, *J. Natl. Cancer Inst.,* 53, 1487, 1974.
42. Hibbs, J. B., Jr., Chapman, H. A., Sr., and Weinberg, J. B., The macrophage as an antineoplastic surveillance cell: biological perspectives, *J. Reticuloendothel. Soc.,* 24, 549, 1978.
43. Nash, J. R. G., Macrophages in human tumors: an immunohistochemical study, *J. Pathol.,* 136, 73, 1982.
44. Narmann, S. J. and Cornelius, J., Concurrent depression of tumor macrophage infiltration and systemic infiltration by progressive cancer growth, *Cancer Res.,* 38, 3453, 1978.
45. Peri, G., Biondi, A., Bottazzi, B., Polentarutti, N., Bodignon, C., and Montovani, A., Tumoricidal and immunoregulatory activity of macrophages from human ovarian carcinomas, *Adv. Exp. Med. Biol.,* 155, 737, 1982.
46. Talmadge, J. E., Key, M., and Fidler, I. J., Macrophage content of metastatic and non-metastatic rodent neoplasms, *J. Immunol.,* 126, 2245, 1981.
47. Wood, G. W. and Gallahon, K. A., Detection and quantitation of macrophage infiltration into primary human tumors with the use of cell surface markers, *J. Natl. Cancer Inst.,* 59, 1081, 1977.
48. Siemann, D. W., Lord, E. M., Keng, P. C., and Wheeler, K. T., Cell subpopulations dispersed from solid tumours and separated by centrifugal elutriation, *Br. J. Cancer,* 44, 100, 1981.
49. Fidler, I. J. and Hart, I. J., Biological and experimental consequences of the zonal composition of solid tumors, *Cancer Res.,* 41, 3266, 1981.

50. Twentyman, P. R. and Yuhas, J. M., Use of a bacterial neutral protease for disaggregation of mouse tumours and multicellular tumour spheroids, *Cancer Lett.,* 9, 255, 1980.
51. Grdina, D. J., Linde, S., and Mason, K., Response of selected tumor cell populations separated from a fibrosarcoma following irradiation *in situ* with fast neutrons, *Br. J. Radiol.,* 51, 291, 1978.
52. Grdina, D. J., Hittelman, W. N., White, R. A., and Meistrich, M. L., Relevance of density, size, and DNA content of tumour cells to the lung colony assay, *Br. J. Cancer,* 36, 659, 1972.
53. Grdina, D. J., Milas, L., Mason, K. A., and Withers, H. R., Separation of cells from a fibrosarcoma in Renograffin density gradients, *J. Natl. Cancer Inst.,* 52, 253, 1974.
54. Grdina, D. J., Basic, I., Mason, K. A., and Withers, H. R., Radiation response of clonogenic cell populations isolated from a fibrosarcoma, *Radiat. Res.,* 63, 483, 1975.
55. Grdina, D. J., Basic, I., Guzzino, S., and Mason, K. A., Radiation response of cell populations irradiated *in situ* and separated from a fibrosarcoma, *Radiat. Res.,* 66, 634, 1976.
56. Grdina, D. J., Sigdestad, C. P., and Jovonovich, J. A., The effect of misonidazole *in situ* on the radiation response of selected tumor subpopulations, *Cancer Clin. Trials,* 3, 149, 1980.
57. Brock, W. A., Swartzendruber, D. E., and Grdina, D. J., Kinetic heterogeneity in density-separated murine fibrosarcoma subpopulations, *Cancer Res.,* 42, 4999, 1982.
58. Durand, R. E., Isolation of cell subpopulations from *in vitro* tumor models according to sedimentation velocity, *Cancer Res.,* 35, 1295, 1975.
59. Miller, R. G. and Phillips, R. A., Separation of cells by velocity sedimentation, *J. Cell. Physiol.,* 73, 191, 1969.
60. Sutherland, R. M., McCredie, J. A., and Inch, W. R., Growth of multi-cell spheroids in tissue culture as a model of modular carcinomas, *J. Natl. Cancer Inst.,* 46, 113, 1971.
61. Howard, A. and Pelc, S. R., Nuclear incorporation of ^{32}P as demonstrated by autoradiographs, *Exp. Cell Res.,* 2, 178, 1951.
62. Heppner, G. H., Dexter, D. L., DeNucci, T., Miller, F. R., and Calabresi, P., Heterogeneity in drug sensitivity among tumor cell subpopulations of a single mammary tumor, *Cancer Res.,* 38, 3758, 1978.
63. Dexter, D. L., Kowalski, H. M., Blazar, B. A., Fligiel, Z., Vogel, R., and Heppner, G. H., Heterogeneity of tumor cells from a single mouse mammary tumor, *Cancer Res.,* 38, 3174, 1978.
64. Heppner, G. H., The challenge of tumor heterogeneity, in *Commentaries on Research in Breast Cancer,* Vol. 1, Bulbrook, R. D. and Taylor, D. J., Eds., Alan R. Liss, New York, 1979, 177.
65. Darzynkywicz, Z., Traganos, F., Andruff, M. L., Sharpless, T., and Melamed, M., Different sensitivity of chromatin to acid denaturation in quiescent and cycling cells as revealed by flow cytometry, *J. Histochem. Cytochem.,* 27, 478, 1979.
66. Traganos, F., Drazynkywicz, Z., Sharpless, J., and Melamed, M. R., Simultaneous staining of ribonucleic and deoxyribonucleic acids in unfixed cells using acridine orange in a flow cytometric system, *J. Histochem. Cytochem.,* 25, 45, 1977.
67. Bauer, K. D., Keng, P. C., and Sutherland, R. M., Isolation of quiescent cells from multicellular spheroids using centrifugal elutriation, *Cancer Res.,* 42, 72, 1982.
68. Wallen, C. A., Higashikubo, R., and Dethlefsen, L. A., Murine mammary tumour cells *in vitro.* I. The development of a quiescent state, *Cell Tissue Kinet.,* 17, 65, 1984.
69. Wallen, C. A., Higashikubo, R., and Dethlefsen, L. A., Murine mammary tumour cells *in vitro.* II. Recruitment of quiescent cells, *Cell Tissue Kinet.,* 17, 79, 1984.
70. Suzuki, N., Frapart, M., Grdina, D. J., Meistrich, M. L., and Withers, H. R., Cell cycle dependency of metastatic lung colony formation, *Cancer Res.,* 37, 3690, 1977.
71. Nicolini, C., Nuclear morphometry, quinternary chromatin structure and cell growth (a review), *J. Submicrosc. Cytol.,* 12, 475, 1980.
72. Puck, T., Cyclic AMP, microtubule-microfilament system and cancer, *Proc. Natl. Acad. Sci. U.S.A.,* 74, 4491, 1977.
73. Ramaekers, F. C. S., Beck, H., Vooijs, G. P., and Herman, C. J., Flow-cytometric analysis of mixed cell populations using intermediate filament analysis, *Exp. Cell Res.,* 153, 249, 1984.
74. Sklarew, R. M., Cytokinetics of subpopulations in mixed heteroploid tumors by television imaging. I. Deconvolution of the S-phase DNA ploidy composition. II. Analysis of the S-phase emptying profile of ploidy subpopulations, *J. Histochem. Cytochem.,* 32, 413, 1984.
75. Sklarew, R. J., Cytokinetics of subpopulations in mixed polyploid tumors by television imaging. III. ^{3}H-Thymidine incorporation by ploidy subpopulations — control of ^{3}H-absorption and emulsion efficiency in autoradiography, *J. Histochem. Cytochem.,* 32, 421, 1984.
76. Frindel, E., Malaise, E. P., Alpen, E., and Tubiana, M., Kinetics of cell proliferation of an experimental tumour, *Cancer Res.,* 27, 1122, 1967.
77. Hermens, A. F., and Barendsen, G. W., Changes of cell proliferation characteristics in a rat rhabdomyosarcoma before and after X-irradiation, *Eur. J. Cancer,* 5, 173, 1969.

78. Watson, J. V., The cell proliferation kinetics of the EMT6/M-AC mouse tumour at four volumes during unperturbed growth *in vivo, Cell Tissue Kinet.,* 9, 147, 1976.
79. Denekamp, J., The cellular proliferation kinetics of animal tumors, *Cancer Res.,* 30, 393, 1970.
80. Humphries, J. E. and Isaacs, J. T., Unusual androgen sensitivity of the androgen-independent Dunning R-3327-G rat prostatic adenocarcinoma: androgen effect on tumor cell loss, *Cancer Res.,* 42, 3148, 1982.
81. Isaacs, J. T. and Coffey, D. S., Adaptation *versus* selection as the mechanism responsible for the relapse of prostatic cancer to androgen ablation as studied in the Dunning R-3327-H, adenocarcinoma, *Cancer Res.,* 41, 5070, 1981.
82. Isaacs, J. T., Heston, W. D. W., Weissman, R. M., and Coffey, D. S., Animal models of the hormone-sensitive and -insensitive prostatic adenocarcinomas Dunning R-3327-H, R-3327-HI, and R-3327-AT, *Cancer Res.,* 38, 4343, 1978.
83. Barlogie, B., Abnormal cellular DNA content as a marker of neoplasia, *Eur. J. Clin. Oncol.,* 20, 1123, 1984.
84. Frankfurt, O. S., Greco, W. R., Slocum, H. K., Arbuck, S. G., Gammara, M., Pavelic, Z. P., and Rustum, Y. M., Proliferative characteristics of primary and metastatic human solid tumors by DNA flow cytometry, *Cytometry,* 5, 629, 1984.
85. Nowell, P. C., Genetic instability in cancer cells: relationship to tumor cell heterogeneity, in *Tumor Cell Heterogeneity: Origins and Implications,* Owens, A. H., Jr., Coffey, D. S., and Baylin, S. B., Eds., Academic Press, New York, 1982, 351.
86. Sandberg, A. A., Chromosomal changes in human cancers: specificity and heterogeneity, in *Tumor Cell Heterogeneity: Origins and Implications,* Owens, A. H., Jr., Coffey, D. S., and Baylin, S. B., Eds., Academic Press, New York, 1982, 370.
87. Cifone, M. and Fidler, I. J., Increasing metastatic potential is associated with increasing genetic instability of clones isolated from murine neoplasms, *Proc. Natl. Acad. Sci. U.S.A.,* 78, 6949, 1981.
88. Johnson, T. S., Raju, M. R., Giltinan, R. K., and Gillette, E. L., Ploidy and DNA distribution analysis of spontaneous dog tumors by flow cytometry, *Cancer Res.,* 41, 3005, 1981.
89. Atkin, N. B., Mattinson, G., and Baker, M. C., A comparison of the DNA content and chromosome number of fifty human tumors, *Br. J. Cancer,* 20, 87, 1966.
90. Barlogie, B., Drewinko, B., Schumann, Jr., Gohde, W., Dosik, G., Latreille, J., Johnston, D. A., and Freireich, E. J., Cellular DNA content as a marker of human neoplasia, *Am. J. Med.,* 69, 195, 1980.
91. Leuchtenberger, C., Leuchtenberger, R., and Davis, M. A., A microspectrophotometric study of the deoxyribonucleic acid (DNA) content in cells of normal and malignant human tissues, *Am. J. Pathol.,* 30, 65, 1954.
92. Miles, C., Chromosome analysis of solid tumors. I. Twenty-eight nonepithelial tumors, *Cancer,* 20, 1253, 1967.
93. Miles, C., Chromosome analysis of solid tumors. II. Twenty-six epithelial tumors, *Cancer,* 20, 1274, 1967.
94. Patek, E., Johannison, E., Krauer, F., and Riotton, G., Microfluorometric grading of mammary tumors, *Anal. Quant. Cytol.,* 2, 264, 1980.
95. Frankfurt, O. S., Slocum, H. K., Rustum, Y. M., Arbuck, S. G., Pavelic, Z. P., Petrelli. N., Huben, R. P., Pontes, E. J., and Greco, W. R., Flow cytometric analysis of DNA aneuploidy in primary and metastatic human solid tumors, *Cytometry,* 5, 71, 1984.
96. Jakobsen, A., Bichel, P., and Sell, A., Flow cytometric investigations of human bladder carcinoma compared to histological classification, *Urol. Res.,* 7, 109, 1979.
97. Tribukait, B., Gustafson, H., and Esposti, P., Ploidy and proliferation in human bladder tumors as measured by flow cytofluorometric DNA — analysis and its relation to histopathology and cytology, *Cancer,* 43, 1742, 1979.
98. Friedlander, M. L., Hedley, D. W., Taylor, I. W., Russell, P., Coates, A. S., and Tattersall, M. H. N., Influence of celluar DNA content on survival in advanced ovarian cancer, *Cancer Res.,* 44, 397, 1984.
99. Lee, S. H., Cancer cell estrogen receptor of human mammary carcinoma, *Cancer,* 44, 12, 1979.
100. Tribukait, B., Hammarbert, C., and Rubio, C., Ploidy, and proliferation patterns in colo-rectal adenocarcinoma related to Dukes' classification and to histopathological differentiation: a flow cytometric study, *Acta Pathol. Microbiol. Immunol. Scand. Sect. A,* 91, 89, 1983.
101. Heliö, H., Karaharju, E., and Nordling, S., Flow cytometric determination of DNA content in malignant and benign bone tumors, *Cytometry,* 6, 165, 1985.
102. Auer, G. V., Fallenius, A. G., Erhardt, K. Y., and Sundelin, B. S. B., Progression in mammary adenocarcinomas as reflected by nuclear DNA content, *Cytometry,* 5, 420, 1984.
103. Straus, M. J., Growth characteristics of lung cancer, in *Lung Cancer,* Straus, J. T., Ed., Grune & Stratton, New York, 1976, 19.

104. Vindeløv, L. V., Flow microfluorimetric analysis of nuclear DNA in cells from solid tumors and cell suspensions, *Virchows Arch. B. Cell Pathol.*, 24, 227, 1977.
105. Baisch, H., Beck, H. P., Christensen, I. J., Hartmann, N. R., Fried, J., Dean, P. N., Gray, J. W., Jett, J. H., Johnston, D. A., White, R. A., Nicolini, C., Zeitz, S., and Watson, J. V., A comparison of mathematical methods for the analysis of DNA histograms obtained by flow cytometry, *Cell Tissue Kinet.*, 15, 235, 1982.
106. Dean, P. N. and Jett, J. H., Mathematical analysis of DNA distributions derived from flow microfluorometry, *J. Cell Biol.*, 60, 523, 1974.
107. Fried, J., Method for the quantitative evaluation of data from flow microfluorometry, *Comput. Biomed. Res.*, 9, 263, 1976.
108. Gray, J. W., Cell cycle analysis of DNA histograms from asynchronous and synchronous cell populations, in *3rd Int. Symp. Pulse Cytophotometry*, Lutz, D., Ed., European Press, Ghent, 1978, 99.
109. Kim, S. and Perry, S., Mathematical methods for determining cell DNA synthesis rate and age distribution utilizing flow microfluorometry, *J. Theor. Biol.*, 68, 27, 1977.
110. Salzman, G. C., Hiebert, R. D., and Crowell, J. H., Data acquisition and display for a high speed cell sorter, *Comput. Biomed. Res.*, 11, 77, 1978.
111. Shackney, S. E., Interrelationships among DNA content distribution, cell kinetics, and cell morphology: theoretical considerations, experimental correlations, and clinical implications, in *Growth Kinetics and Biochemical Regulation of Normal and Malignant Cells*, Drewinko, B., and Humphrey, R. M., Eds., Williams & Wilkins, Baltimore, 1977, 391.
112. Schuette, W. H., Shackney, S. E., MacCollum, M. A., and Smith, C. A., High resolution method for the analysis of DNA histograms that is suitable for the detection of multiple aneuploid G_1 peaks in clinical samples, *Cytometry*, 3, 376, 1983.
113. Ritch, P. S., Shackney, S. E., Schuette, W. H., Talbot, T. L., and Smith, C. A., A practical graphical method for estimating the fraction of cells in S in DNA histograms from clinical tumor samples containing aneuploid cell populations, *Cytometry*, 4, 66, 1983.
114. Shapiro, J. R., Yung, W.-K. A., and Shapiro, W. R., Isolation, karyotype and clonal growth of heterogeneous subpopulations of human malignant gliomas, *Cancer Res.*, 41, 2349, 1981.
115. Valeriote, F. and Van Putten, L., Proliferation-dependent cytotoxicity of anticancer agents: a review, *Cancer Res.*, 35, 2619, 1975.
116. Mendelsohn, M. L., Autoradiographic analysis of cell proliferation in spontaneous breast cancer of C_3H mouse. III. Growth fraction, *J. Natl. Cancer Inst.*, 28, 1015, 1962.
117. Dethlefsen, L. A., In quest of the quaint quiescent cells, in *Radiation Biology in Cancer Research*, Meyn, R. E. and Withers, H. R., Eds., Raven Press, New York, 1980, 415.
118. Bahler, D. W., Lord, E. M., Kennel, S. J., and Horan, P. K., Heterogeneity and clonal variation related to cell surface expression of a mouse lung-tumor-associated antigen quantified using flow cytometry, *Cancer Res.*, 144, 3317, 1984.

Chapter 6

ENVIRONMENTAL ASPECTS OF TUMOR HETEROGENEITY

"At any time during progression the number of subpopulations present in a tumor, and the extent of their phenotypic diversity, will reflect the selection pressures encountered during the lifetime of the tumor."[1]

I. INTRODUCTION

This statement by Poste et al.[1] in 1984 admirably reflects the interdependence between the phenotypic diversity of tumor subpopulations and the intimate relationship the expression of such diversity must have to the environment within which these subpopulations reside. While aspects of intrinsic cellular tumor heterogeneity have been discussed in Chapters 1 and 2, and various quantitative aspects of heterogeneity in Chapter 5, all aspects of intrinsic tumor cell expression are modulated (albeit perhaps to different extents), by local microenvironmental influences.[2-4] It is the purpose of this chapter to discuss some of the environmental influences that impact directly or indirectly upon intraneoplastic diversity.

II. TUMOR HYPOXIA

A major microenvironmental influence upon solid tumors lies in the finding that a significant fraction of the growing neoplasm may consist of hypoxic cells. This has important clinical implications, particularly in the use of ionizing radiation in the treatment of cancer (as will be discussed in Chapter 8), as such hypoxic cells are significantly more resistant to this cytotoxic modality. The initial demonstration of hypoxic areas in tumors occurred in 1955 when Thomlinson and Gray[5] histologically showed that in human bronchial carcinoma, within which malignant cells grow in cords, any cord with a diameter of more than about 0.4 mm contained a central area of necrotic tissue. It was then shown that this distance related well to calculations of the diffusibility of molecular oxygen within actively respiring tissue. A model was developed in which solid tumors contained a necrotic (anoxic) core, surrounded by a zone of hypoxic cells, with a rim of actively proliferating (oxic) cells (Figure 1). The key to this model was the fact that as oxygen and carbon dioxide exchange would occur at the capillary/tumor cell interface, a gradient in oxygen availability at increasing distances from the capillary (due to cellular metabolic demands) would be encountered. It should be emphasized that while hypoxic cells may not be actively proliferating, they may retain clonogenic ability (Chapter 5). Therefore, as cells that are potentially able to proliferate (given reexposure to oxygen), they are a significant factor in the ultimate curability of cancers.

Additional evidence of the existence of hypoxic cells in human cancers has been derived from direct measurements of lowered oxygen tension in neoplasms,[6,7] a direct correlation between lowered blood hemoglobin levels and the control of localized carcinoma of the cervix,[8] some improvement in control rates of malignancies when patients were given hyperbaric oxygen during irradiation,[9] and some success when specific chemical agents (termed hypoxic cell sensitizers) were used in conjunction with localized tumor irradiation.[10] All of these studies point to oxygen as being a prime factor in the internal cellular state of the tumor.

There are several important features that relate to the presence of hypoxic cells within (heterogeneous) cancers. First, it should be appreciated that there is *time-de-*

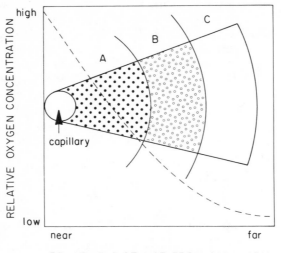

FIGURE 1. Illustration of the diffusion of oxygen from a capillary in a solid tumor. Due to the metabolic use of oxygen by the tumor cells, the concentration of oxygen will decrease with increasing distance away from the capillary. Three regions can be distinguished: (A), a region within approximately 150 μm from the capillary containing oxygenated cells, which are usually actively proliferating: (B), an intermediate region of hypoxic cells which typically show decreased cell proliferation although they are capable of continuing proliferation if they are reexposed to oxygen (note: these cells are resistant to ionizing radiation); and (C), a region of anoxic, usually necrotic cells.

pendent heterogeneity in the development of tumor hypoxia.[11] At small sizes, the neoplasm can be totally supplied with oxygen by simple diffusion; the theoretical limit to this would be about 0.15 to 0.20 mm in radius. However, even at this small size, given an average diameter for a single cell (e.g., 10 μm), the neoplasm could contain easily 20,000 cells. Even if the tumor at this size contained a large fraction of infiltrating host cells (e.g., 50%),[12] there are still about 10,000 neoplastic cells that are actively proliferating. If the mutation rate of such cells is high,[13] this suggests that subpopulations will be present even at the earliest times in the evolution of the tumor. However, imposed on this early cellular heterogeneity is the environmental heterogeneity produced as the cancer outgrows its source of oxygen. *Therefore, both cellular and environmental heterogeneity exist almost from the beginning of the evolution of solid tumors.* Once a state of internal hypoxia is produced, the available evidence suggests that this level will remain essentially constant with further growth.[14] This level will change drastically with therapeutic intervention, however, due to the fact that oxic cells are killed preferentially, because of their position in proximity to capillaries where oxygen and/or drug concentrations will be at their highest. As a consequence, immediately after treatment there will be a relative increase in the percentage of hypoxic (radiation resistant but clonogenic) cancer cells relative to the total cell population. This situation does not persist, however, due to a process termed reoxygenation, in which hypoxic tumor cells become oxygenated (a phenomenon at least partially due to the removal of treatment-killed oxic cells), which reestablishes the pretreatment level of hypoxia.[14-16] Therefore, it has been stated that the extent of hypoxia appears to be a biological constant for a given neoplasm. However, a recent report by Dorie and Kallman[17] who studied the kinetics of reoxygenation in the RIF-1 tumor showed that for at least 2 weeks after X-irradiation the fraction of hypoxic cells remained about ten-fold greater than that existing before irradiation. The authors suggest that this result may be due to effects on stromal tissue, although it is possible that it is in some way related to the fact that the RIF-1 is itself a heterogeneous tumor. It is important to also note that there exists marked *intertumor heterogeneity with regard to hypoxia*,[18,19] and that the percentage of hypoxic cells can range from essentially none to about 80%.[18-21] The origins of the neoplasm have an important relationship to the impact of the environment, and a typical value for the "average" tumor would be about 15 to 20% hypoxic cells.[18]

Herndon et al.[22] have performed an interesting experiment in which mice with mammary tumors were exposed to varying atmospheric pressures ranging from 0.33 to 2 atm for long periods of time. The composition of the breathing gases also varied, including air, 100% oxygen, and other nitrogen-oxygen mixtures. The work of Herndon et al.[22] was designed to separate the variables of pressure vs. gas content per se (i.e., pressure effects on neoplastic growth vs. hypoxia[23]). This situation would qualify as an investigation of "selection pressure" as described by Poste et al.[1] It was found that tumor growth decreased as the pressure decreased, and that this response could be offset by increasing the percentage of oxygen in the breathing mixture. This is an indication that it is the lack of oxygen, not pressure per se, that is the driving force in the induction of hypoxia which is then accompanied by a decreased proliferation rate.

While *inter* tumor environmental heterogeneity (with respect to hypoxia) has been clearly demonstrated, there is very little information on *intra* tumor heterogeneity. Leith et al.[24] have recently demonstrated that subpopulations of cells derived from the human colon adenocarcinoma DLD-1, when grown as xenograft tumors in nude mice, varied significantly in their hypoxic fractions. The two main subpopulations termed clones A and D, were shown to produce cancers containing approximately 3% and essentially 0% hypoxic cells, respectively. This is the first experimental evidence of such diversity in the environmental influence upon subpopulations of cells from a heterogeneous human tumor and indicates that differential hypoxia in human neoplasms should not be unexpected. Thus, diversity due to extrinsic influences may be superimposed upon a pattern of intrinsic cellular heterogeneity. This work is also discussed in Chapter 8. This finding points up one of the fundamental problems associated with the strategy of treatment of heterogeneous tumors, namely the differential expression of characteristics.[25] In this case, differential expression of radiation sensitivity is directly a result of environmental modulations, and suggests that different clonal subpopulations may be hypoxic within the solid tumor in an unpredictable manner. This uncertainty will influence the outcome of therapy protocols (Chapters 8 and 9).

III. TUMOR pH

It is important to consider the effects of differing pH within a solid tumor, as this has important therapeutic implications. Many neoplasms appear to exist at a lower pH (i.e., increased acidity) than does surrounding normal tissue.[26-30] An important therapeutic consequence is that the lowered pH could sensitize cells to at least one type of cytotoxic treatment, that being hyperthermia (i.e., temperatures of 40 to 45°C) (cf., Chapter 8). Biochemically, the acidity of a tumor is a result of anaerobic metabolism in hypoxic cells with the production of 2 molecules of lactate per glucose molecule. In oxic cells, glucose metabolism produces six acids via glycolysis and tricarboxcyclic acid cycle. Metabolic acidity is also produced by the pentose shunt (one acid molecule for every glucose). The main buffering system in mammalian cells is the CO_2/ HCO_3^- system, which is constantly in dynamic flux and is affected by the activity of carbonic anhydrase,[31] and cyclic nucleotides.[32,33] While CO_2 will diffuse out of the cell unless it is hydrated to form HCO_3^-, lactate is primarily transported out of the cell by carrier-mediated diffusion which is pH dependent.[32,33]

It is important to distinguish between the extracellular pH of tumors (pH_e) and the intracellular pH (pH_i). Although there is much evidence that the pH_e of neoplasms is slightly more acidic than that of the surrounding normal tissues,[34-38] the pH_i appears to be similar to normal tissues.[39-41] Kitagawa and Kuroiwa[42] have demonstrated that in the process of induced carcinogenesis in rat liver, at no time was the pH_i of liver tumors different from normal liver values. Spencer and Lehninger[35] have demonstrated that Ehrlich ascites cells are able to maintain a more alkaline pH_i over a wider range of extracellular pH than do normal cells. This ability declines as the external lactic acid level rises. Tumor cells may contain mechanisms to obviate intracellular acidification by metabolism,[43] a feature that may be dependent on vascularization. To our knowledge, no studies have yet been performed on the relative abilities of subpopulations of cancer cells to maintain their pH_i. Further, no studies have been performed on biochemical indexes of glycolysis such as the Pasteur effect or the Warburg effect in tumor subpopulations. Such studies should have important implications for the control of cell growth in solid neoplasms. For example, cells have both intra- and extracellular pH optima for growth, and stimulation to enter proliferation can occur with a shift to a more alkaline condition.[44] Taylor and Hodson[45] have also shown that there may be a pH-dependent restriction point in the G_1 phase of tumor cells.

Although few data are available on pH gradients in heterogeneous neoplasms, there is little reason to doubt that gradients will exist given the zonal composition of tumors, and as a function of distance from capillaries. In this regard, Thistlewaite et al.[46] have investigated pH_e distributions in human tumors and found that the average pH was 6.8. They also noted that there appeared to be no significant differences among tumors of different histologies. Another recent finding bearing indirectly on this question is the demonstration by Leith[47] that exposure of two subpopulations from the DLD-1 human colon adenocarcinoma to external acidities of either 7.40 or 6.75 would *differentially* sensitize the subpopulations to killing at increased temperatures (c.f., Chapter 8). These data indicate that significant differences exist among malignant subpopulations in their respective abilities to successfully respond to changes in pH (when combined with hyperthermia). Still more work needs to be done in this area.

IV. NUTRIENTS

There have been many attempts at modeling the growth of solid tumors based on considerations of diffusion of nutrient substances from the capillary bed to the cells, as active cell proliferation is strongly dependent on distance away from nutrient-supplying blood vessels. While much effort has gone into the description of gaseous diffusion (e.g., oxygen and carbon dioxide), somewhat less has gone into transfer of substances such as glucose. The specific data needed to correlate tumor cell growth in vivo with availability of specific nutrients is largely lacking.

The uncertainty of nutrient supply raises a correlated question dealing with the vascular bed. It is known that the vascular supply and the actual blood volumes vary among different types of neoplasms. It is not known whether the blood supplies of various areas of zonal tumors might also vary according to the cellular phenotype. As cancer cells differ in such parameters as cell volume, motility, proliferative capacity, etc. (tumor angiogenic factor expression?), it would not be surprising if the blood volume of heterogeneous cancers also expressed some degree of internal local variation.

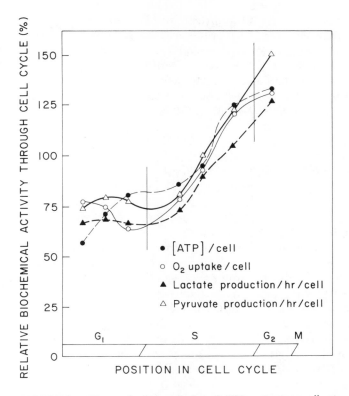

FIGURE 2. Changes in the expression of ATP content per cell, oxygen uptake per cell, and lactate and pyruvate production per hour per cell, as a function of position in the cell cycle for Ehrlich ascites tumor cells. The data have been normalized to the activities of asynchronously growing tumor cells.

There is strong correlation between the observation of necrotic areas within neoplasms and the diffusibility of glucose,[48,49] as by the time a size of about 2-mm diameter is reached, the intratumor glucose drops essentially to zero. Therefore, there is a relationship between glucose and oxygen availability and the development of necrotic and hypoxic areas. Closely related to the utilization of glucose and its potential role as a limiting agent in cell proliferation and resultant phenotypic expression is the overall energy metabolism and use of ATP through the cell cycle of tumor cells. Skog et al.[50] have used centrifugal elutriation to separate cells through the cell cycle (c.f., Chapter 5), and oxygen consumption, and lactate, pyruvate, and ATP use were determined with cell cycle distribution of separated cells confirmed by flow cytometry (Chapter 5). In Figure 2, several biochemical parameters through the cell cycle are shown, and significant changes occur, with parameters being at their lowest in G_1, and then increasing through S phase and into G_2 and M. The authors suggested that the total energy requirement for a tumor cell to traverse the cell cycle was likely directly related to cell volume, with larger cells having a larger ATP requirement. In situations of nutrient deprivation, this situation might tend to select against the larger cells of a heterogeneous cancer. In this regard, significant differences in the cell volumes of subpopulations have been reported in the heterogeneous DLD-1 human colon adenocarcinoma.[51]

Recent observations by Wallen et al.,[52,53] previously described in Chapter 5, are also relevant here. In the studies of Wallen et al., different subpopulations from the mouse mammary tumor system developed by Dexter et al.[54] were forced into a quiescent state of proliferation in vitro by nutrient deprivation. This deprived state was associated with a greatly reduced level of intracellular RNA in all subpopulations. However, in this starved condition, the clonal lines differed significantly in their clonogenicity, and also in the rate at which they returned to the proliferative state when exposed to fresh nutrients. Both of these findings are important in terms of the effects of environmental selection pressures upon the relative survival and potential response to cell cycle specific cytotoxic agents in heterogeneous cancers.

V. OTHER

Other agents that might influence the local microenvironment within a solid tumor would be such factors as the host cell percentage (discussed in Chapter 7); hormonal influences,[55-57] the presence of biochemical agents produced by the host which might mediate the state of cancer cell differentiation (e.g., prostaglandins, vitamin A), and other as yet undefined "selection pressures".

It is relevant to comment upon differences in the microenvironment of tumors vs. normal tissues as indicated by nuclear magnetic resonance (NMR) studies on water content. By application of a resonating magnetic field to a biological sample, atomic nuclei will absorb energy, and radiofrequency currents will be induced which decay ("relax") with specific time constants. These parameters indicate the freedom of movement of water molecules in the cell and mirror subtle changes in cell physiology. Changes in signals reflect molecular aspects of membranes, cytoskeleton, and cellular organelles, and also provide an index of cell transformation. Specifically, it has been found that NMR relaxation times are significantly longer in neoplastic tissue than in normal tissues. On the average, in 19 measurements on human neoplastic tissues vs. their counterpart untransformed tissues, tumor values were about 1.5 times longer. Other studies have shown that it is possible to distinguish normal mammary tissue, hyperplastic alveolar nodules, and overt mammary cancers using these parameters, illustrating that the physiology of intracellular water state correlates strikingly well with the progression of tumors to more malignant states.[58] Butel et al.[59] have shown that NMR parameters are also different among established mouse mammary cancer lines in vitro, and point out that these differences among mouse mammary tumor lines are correlated with marked phenotypic heterogeneity. Therefore, NMR signals obtained from tumor tissue may provide an independent, quantitative physical index of intraneoplastic diversity, although this has yet to be studied systematically.

VI. SUMMARY AND CONCLUSIONS

While the initial supposition of this chapter that environmental "selection pressures" exist and modulate tumor cell phenotype is quite valid, it is clear that such pressures are not well described. In the area of tumor cell heterogeneity, to our knowledge there are only a few publications relating to the way in which agents such as molecular oxygen, pH, or nutrient deprivation might change the evolution of subpopulations of cancer cells.[24,47,52,53] The results of such environmental agents will be expressed pleiotropically in a heterogeneous solid tumor, having effects on cell proliferation, response to therapeutic agents, etc. A major conclusion of this chapter must be that much further work on the way environmental agents modify cellular biochemistry, potential cell interactions, cellular kinetics, etc., is necessary. Also, studies specifically addressing effects on human cancer cells need to be performed. A final note of caution is needed because metastases in different locations in a patient will be exposed to different extents of environmental modulation, which may impact on therapeutic responses.

REFERENCES

1. Poste, G., Greig, R., Tzeng, J., Koestler, T., and Corwin, S., Interactions between tumor cell subpopulations in malignant tumors, in *Cancer Invasion and Metastasis: Biologic and Therapeutic Aspects*, Nicolson, G. L. and Milas, L., Eds., Raven Press, New York, 1984, 226.
2. Schirrmacher, V., Shifts in tumor cell phenotypes induced by signals from the microenvironment: relevance for the immunobiology of cancer metastasis, *Immunobiology*, 157, 89, 1980.
3. Nowell, P. C., The clonal evolution of tumor cell populations, *Science*, 194, 23, 1976.
4. Nowell, P. C., Tumors as clonal proliferation, *Virchows Arch. B: (Cell Pathol.)*, 29, 145, 1978.
5. Thomlinson, R. H. and Gray, L. H., The histological structure of some human lung cancers and the possible implications for radiotherapy, *Br. J. Cancer*, 9, 539, 1955.
6. Cater, D. B. and Silver, I. A., Quantitative measurements of oxygen tension in normal tissues and in the tumors of patients before and after radiotherapy, *Acta Radiol.*, 53, 233, 1960.
7. Evans, N. T. S. and Naylor, P. F. D., The effect of oxygen breathing and radiotherapy upon the tissue oxygen tension of some human tumours, *Br. J. Cancer*, 36, 418, 1963.
8. Bush, R. S., Jenkin, R. D. T., Allt, W. E. C., Beale, F. A., Bean, H., Dembo, A. J., and Pringle, J. F., Definitive evidence for hypoxic cells influencing cure in cancer therapy, *Br. J. Cancer*, 37 (Suppl. III), 302, 1978.
9. Henk, J. M., Kunkler, P. B., and Smith, C. W., Radiotherapy and hyperbaric oxygen in head and neck cancer — Final report of first controlled clinical trail, *Lancet*, II, 101, 1977.
10. Urtasun, R., Band, P., Chapman, J. D., Feldstein, M. L., Mielke, B., and Freyer, C., Radiation and high-dose metronidazole in supratentorial glioblastomas, *N. Eng. J. Med.*, 294, 1364, 1976.
11. Thomlinson, R. H., Changes of oxygenation in tumors in relation to irradiation, *Front. Radiat. Ther. Oncol.*, 3, 109, 1968.
12. Blazar, B. A. and Heppner, G. H., In situ lymphoid cells of mouse mammary tumors. II. The characterization of lymphoid cells separated from mouse mammary tumors, *J. Immunol.*, 120, 1881, 1978.
13. Goldie, J. H. and Coldman, A. J., A mathematical model for relating the drug sensitivity of tumors to their spontaneous mutation rate, *Cancer Treat. Rep.*, 63, 1727, 1979.
14. Kallman, R. F., The phenomenon of reoxygenation and its implications for fractionated radiotherapy, *Radiology*, 105, 135, 1972.
15. Kallman, R. F. and Rockwell, S., Effects of radiation on animal tumor models, in *Cancer: A Comprehensive Treatise*, Vol. 6, Becker, F. F., Ed., Plenum Press, New York, 1977, 225.
16. Kennedy, K. A., Teicher, B. A., Rockwell, S., and Sartorelli, A. C., The hypoxic tumor cell: a target for selective cancer chemotherapy, *Biochem. Pharmacol.*, 29, 1, 1980.
17. Dorie, M. J. and Kallman, R. F., Reoxygenation in the RIF-1 tumor, *Int. J. Radiat. Oncol. Biol. Phys.*, 10, 687, 1984.
18. Hall, E. J., Solid tumor systems and reoxygenation, in *Radiobiology for the Radiotherapist*, Harper & Row, Hagerstown, Md., 1978, 223.
19. Rockwell, S., Moulder, J. E., and Martin, D. F., Tumor-to-tumor variability in the hypoxic fractions of experimental rodent tumors, *Radiother. Oncol.*, 2, 57, 1984.
20. Leith, J. T., Schilling, W. A., and Wheeler, K. T., Cellular radiosensitivity of a rat brain tumor, *Cancer*, 35, 1545, 1975.
21. Denekamp, J., Hirst, D. G., Stewart, F. A., and Terry, N. H. A., Is tumor radiosensitization by misonidazole a general phenomena?, *Br. J. Cancer*, 41, 1, 1980.
22. Herndon, B. L., Lally, J. J., and Hacker, M., Atmospheric pressure effects on tumor growth: hypobaric anoxia and growth of a murine transplantable tumor, *J. Natl. Cancer Inst.*, 70, 739, 1983.
23. Mori-Chavez, P., Upton, A. C., Salazar, M., Jr., and Conklin, W. J., Influence of altitude on late effects of radiation in RF/Un mice: observations on survival time, blood changes, body weight, and incidence of neoplasms, *Cancer Res.*, 30, 913, 1970.
24. Leith, J. T., Bliven, S. F., Lee, E. S., Glicksman, A. S., and Dexter, D. L., Intrinsic and extrinsic heterogeneity in the responses of parent and clonal human colon carcinoma xenografts of X-irradiation, *Cancer Res.*, 44, 3757, 1984.
25. Foulds, L., *Neoplastic Development*, Vol. 2, Academic Press, New York, 1975.
26. Overgaard, J. and Overgaard, K., Effect of environmental acidity on the hyperthermic treatment of tumor cells, *IRCS Med. Sci. Biochem.* (Cancer: Cell and Membrane Biology), 3, 386, 1972.
27. Freeman, M. L., Dewey, W. C., and Hopwood, L. E., Effect of pH on hyperthermic cell survival: brief communication, *J. Natl. Cancer Inst.*, 58, 1837, 1977.
28. Gerweck, L. and Rottinger, E., Enhancement of mammalian cell sensitivity to hyperthermia by pH alteration, *Radiat. Res.*, 67, 508, 1976.
29. Gullino, P. M., Grantham, F. H., Smith, S. A., and Haggerty, A. C., Modifications of the acid-base status of the internal milieu of tumors, *J. Natl. Cancer Inst.*, 34, 857, 1965.

30. Warburg, O., Über den Stoffwechsel der Tumoren, Springer-Verlag, Berlin, 1926.
31. Thomas, R. C., The effect of carbon dixoide on the intracellular pH and buffering power of snail neurons, *J. Physiol. (London)*, 255, 713, 1976.
32. Fenton, R. A., Gonzales, N. C., and Clancy, R. L., The effect of dibutyryl cAMP and glucagon on the myocardial cell pH, *Respir. Physiol.*, 32, 213, 1978.
33. Boron, W. F., Russell, J. M., Brodwick, M. S., Keifer, D. W., and Roos, A., Influence of cAMP on intracellular pH regulation and chloride fluxes in barnacle muscle fibers, *Nature (London)*, 276, 511, 1978.
34. Belt, J. A., Thomas, J. A., Buchsbaum, R. N., and Racker, E., Inhibition of lactate transport and glycolysis in Ehrlich ascites tumor cells by bioflavinoids, *Biochemistry*, 18, 3506, 1979.
35. Spencer, T. L., and Lehninger, A. L., L-Lactate transport in Ehrlich ascites tumor cells, *Biochem. J.*, 154, 405, 1976.
36. Kahler, H. and Robertson, W., Hydrogen ion concentration of normal liver and hepatic tumors, *J. Natl. Cancer Inst.*, 3, 495, 1943.
37. Naesland, J., Further attempts at direct chemical attack upon malignant tumors, based on the change of the pH after injection of glucose and malonate, *Acta Soc. Med. Ups.*, 60, 150, 1955.
38. Eden, M., Haines, B., and Kahler, H., The pH of rat tumors measured *in vito, J. Natl. Cancer Inst.*, 16, 541, 1955.
39. Poole, D., Butler, T. C., and Williams, M. E., The effects of nigericin, valinomycin, and 2,4-dinitrophenol on intracellular pH, glycolysis and K^+ concentration of Ehrlich ascites tumor cells, *Biochim. Biophys. Acta*, 266, 463, 1972.
40. Schloerb, P. R., Blackburn, J. J., Grantham, D. S., Mallard, D. S., and Cage, G. K., Intracellular pH and buffering capacity of the Walker-256 carcinoma, *Surgery*, 58, 5, 1965.
41. Poole, D., Intracellular pH of the Ehrlich ascites tumor as it is affected by sugars and sugar derivatives, *J. Biol. Chem.*, 242, 3731, 1967.
42. Kitagawa, Y. and Kuroiwa, Y., Change in intracellular pH of rat liver during azo-dye carcinogenesis, *Life Sci.*, 18, 441, 1976.
43. Gillies, R. J., Intracellular pH and growth control in eukaryotic cells, in *The Transformed Cell*, Cameron, I. L., and Pool, T. B., Eds., Academic Press, New York,, 1981, 348.
44. Fodge, D. W. and Rubin, H., Activation of phosphofructokinase by stimulants of cell multiplication, *Nature (London), New Biol.*, 246, 181, 1973.
45. Taylor, I. W. and Hodson, P. J., Cell cycle regulation by environmental pH, *J. Cell Physiol.*, 121, 517, 1984.
46. Thistlewaite, A. J., Leeper, D. B., Moylan, D. J., III, and Nerlinger, R. E., pH distribution in human tumors, *Int. J. Radiat. Oncol. Biol. Phys.*, accepted for publication, 1985.
47. Leith, J. T., Unequal modification of hyperthermic tumor cell killing in parent and clonal cell lines from a heterogeneous human tumor, *Int. J. Radiat. Biol.*, in press, 1985.
48. Li, C. K. N., The glucose distribution in 9L rat brain multicell tumor spheroids and its effect on cell necrosis, *Cancer*, 50, 2066, 1982.
49. Li, C. K. N., The role of glucose in the growth of 9L multicell tumor spheroids, *Cancer*, 50, 2074, 1982.
50. Skog, S., Tribukait, B., and Sundius, G., Energy metabolism and ATP turnover time during the cell cycle of Ehrlich ascites tumour cells, *Exp. Cell Res.*, 23, 1982.
51. Leith, J. T., Bliven, S. F., Arundel. C. M., and Glicksman, A. S., Separation and characterization of cellular subpopulations from a heterogeneous human colon tumor system by centrifugal elutriation, *Proc. Am. Soc. Cancer Res.*, 24, 32, 1983.
52. Wallen, C. A., Higashikubo, R., and Dethlefsen, L. A., Murine mammary tumour cells *in vitro*. I. The development of a quiescent state, *Cell Tissue Kinet.*, 17, 65, 1984.
53. Wallen, C. A., Higashikubo, R., and Dethlefsen, L. A., Murine mammary tumour cells *in vitro*. II. Recruitment of quiescent cells, *Cell Tissue Kinet.*, 17, 79, 1984.
54. Dexter, D. L., Kowalski, H. M., Balzar, B. A., Fligiel, Z., Vogel, R., and Heppner, G. H., Heterogeneity of tumor cells from a single mouse mammary tumor, *Cancer Res.*, 38, 3174, 1978.
55. Humphries, J. E. and Isaacs, J. T., Unusual androgen sensitivity of the androgen-independent Dunning R-3327-G rat prostatic adenocarcinoma: androgen effect on tumor cell loss, *Cancer Res.*, 42, 3148, 1982.
56. Sluyser, M. S. and Hart, G., Calculation of percentage of hormone independent transplantable cells in experimental mammary tumors, *J. Theor. Biol.*, 100, 701, 1983.
57. Pollack, A., Block, N. L., Stover, B. J., and Irvin, G. L., III, Tumor progression in serial passages of the Dunning R3327-G rat prostatic adenocarcinoma: growth rate response to endocrine manipulation, *Cancer Res.*, 45, 1052, 1985.

58. Beall, P. T., Asch, B. B., Medina, D., and Hazlewood, C., Distinction of normal, preneoplastic, and neoplastic mouse mammary cells and tissues by nuclear magnetic resonance techniques, in *The Transformed Cell,* Cameron, I. B. and Pool, T. B., Eds., Academic Press, New York, 1981, 294.
59. Butel, J. S., Dudley, J. P., and Medina, D., Comparison of the growth properties *in vitro* and transplantability of continuous mouse mammary tumor cell lines and clonal derivatives, *Cancer Res.,* 37, 1982, 1977.

Chapter 7

CLONAL INTERACTIONS AMONG TUMOR SUBPOPULATIONS

"Tumors are more complex than a roll call of their subpopulations, however many there are, may suggest."[1]

I. INTRODUCTION

It has been recently pointed out[1-7] that not only are tumors diverse in terms of their cellular composition (e.g., clonal subpopulations), but that there also may exist "interactions" among subpopulations. The basic postulate of such interactive processes is that one subpopulation of cells could modulate the activities of a different subpopulation.[8] For example, such modulation could be expressed in terms of intrinsic growth properties, metastatic potential, or sensitivity to a specific cytotoxic treatment. This basic postulate then conceptually raises the idea of an intratumor ecosystem,[2] with mutual interactions.[3] A corollary to this viewpoint is that introduction of instability into such an ecosystem (e.g., by intervention with a cytotoxic agent) might lead to the "relentless emergence of new subpopulations with enhanced ... capacities."[9]

The concept of clonal interactions is therefore clearly a significant one. Besides relating to the ultimate success or failure of cancer-directed therapy, it invokes the existence of basic biological processes that, if they truly exist, are of fundamental importance. Therefore, this chapter will present the evidence pertaining to the presence of clonal interactions, the mechanisms through which such interactions might be evinced, the potential therapeutic implications of such interactions, and hopefully, experimentation necessary to further explore the phenomenon.

II. REQUIREMENTS FOR OBSERVATION OF CLONAL INTERACTIONS

In order to observe clonal interactions, one much rely on manipulation of phenotypic differences (i.e., the "response phenotype") among tumor subpopulations. Such differences could lie in relative growth rates or cell kinetic behavior, sensitivity to specific cytotoxic modalities, metastatic potential, immunologic properties, etc. Still, it must be appreciated that not only must such differential aspects exist, they must also be amenable to quantitative analysis to provide indication of their "potency". Therefore, clonal interactions must necessarily be operationally defined within the limits and constraints of the experimental systems used. In this regard, it is necessary to describe the experimental model systems that have been used to approach the phenomenon of clonal interactions.

III. IN VIVO EXPERIMENTAL STUDIES

A mouse mammary tumor system has been used extensively by Heppner and colleagues[2,10-18] to study both in vivo and in vitro aspects of clonal interactions. This system has been well characterized,[10-18] as has been its responses to various cytotoxic treatments.[18-21] In this system, one original, spontaneously arising mammary tumor was used to obtain four separate tumor lines in vitro,[17,18] which were termed 68H, 168, 66, and 67. Shortly thereafter, a fifth line (4.10) was obtained from a metastatic lung nodule.

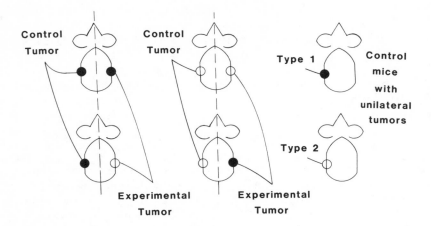

FIGURE 1. Illustration of bilaterally paired solid tumors obtained from tumor cell subpopulations used in in vivo clonal interaction experiments.

For in vivo studies, solid tumors were produced by injecting cells subcutaneously.[11] To monitor clonal interactions, the authors used bilateral paired tumors, which then could be assayed in terms of their relative growth effects upon each other (e.g., 66 plus 67 vs. their respective control paired tumors, 66 plus 66 or 67 plus 67), as illustrated in Figure 1. The endpoints included incidence (i.e., the number of tumors as a percentage of the total number of sites injected), latency (i.e., the average time needed for a tumor to appear), and growth rate (mm²/day).

It was found that the 66 × 66 combination was about 4 to 5 times as inherently tumorigenic as the combination of 67 × 67. However, with the pairing of heterogeneous tumor lines (i.e., 66 plus 67) it was found that line 66 reduced the incidence of line 67 tumors from 63% to 36%.[22,23] Conversely, line 67 did not significantly decrease the incidence of line 66 tumors, although there was a significantly longer latency period (from 35 to 50 days). Therefore, it can be stated that sublines 66 and 67 grew less well when paired together.

Similar experiments were carried out using subpopulations 168 and 410 (the metastatic line). Line 168 cells did not affect any growth parameters of line 410. However, line 410 did reduce slightly the incidence of line 168 tumors. Interestingly, line 410 when injected bilaterally showed "auto" inhibition of the contralateral neoplasm in terms of an increased latency period (16 to 26 days). It was further noted that the inhibitory effect of line 410 cells could be strengthened by allowing time to elapse between the inoculation of line 410 cells and inoculation of either line 410 or 168 cells.

A third combination of subpopulations, lines 68H and 168, provided additional data in that whereas line 68H increased the latency of line 168 carcinomas, the converse experiment (line 168 acting upon line 68H) actually demonstrated an *increase* in the incidence of line 68H tumors (10% to 50%).

In an attempt to unfold the complexities of these experiments, whole body irradiation was given 2 days prior to injection of line 168 and 410 tumor cells, and it was found that the inhibitory effect of line 410 cells was abolished. This indicated an immunological mechanism for the inhibitory effect. This was confirmed by rechallenging mice previously immunized with tumor cells with new tumor cells.[11]

All of the experiments described above indicate that subpopulations from the same tumor can influence the growth of other subpopulations. As stated by Miller et al.,[11] "the behavior of heterogeneous cancers is not necessarily predictable from knowledge of the behavior of their component parts."

Miller et al.[24] have extended this work to include interactions among subpopulations that impinge on sensitivity to cytotoxic agents. In vivo the sensitivities of selected mouse mammary tumor subpopulations (lines 410 and 168) were examined with regard to cyclophosphamide (CY), an alkylating agent used in the therapy of breast cancer. In an initial study, cells were injected and repeated treatments with CY were given at weekly intervals for 1 month and effects on the 410 solid tumor were evaluated. Cyclophosphamide administration did not significantly alter the incidence of line 410 carcinomas when 410 tumors were the bilateral neighbor. However, when line 410 tumors received CY treatment and had line 168 tumors as their bilateral mate, there was a slight decrease in the incidence of line 410 cancers (from 94 to 70%). Therefore, there appeared to be some interactions between cytotoxic drug administration plus the presence of a different, albeit closely related tumor line in terms of inhibition of neoplastic growth. The experiment was repeated in mice that had received whole body γ-ray irradiation to suppress the immune response, and the inhibitory effect of line 168 on line 410 tumors in CY treated mice was still present, indicating that the effect was not produced through immunological mechanisms.

The reverse experiment was also done in which the effects of line 410 tumors on line 168 tumors were studied.[24] It was found that at 35 mg/kg of CY, there was no effect on line 168 tumor incidence. However, when line 410 tumors were on the opposite flank of the mice, a definite interaction was noted; line 168 tumor incidence dropped from 94 to 60%. This decreased incidence was greater than would be expected in control experiments where the effect of 410 tumors on line 168 neoplasms was determined. Therefore, line 410 appeared to exert an additional degree of inhibition of line 168 carcinomas in the presence of CY above and beyond its intrinsic immunosuppressive effect.[25]

In another interesting experiment, Miller et al.[24] used a bioassay in which when tumors has reached a size of about 5-mm diameter, mice were given a range of CY doses. The premise was that a solid tumor burden might influence the extent of in vivo drug activation,[26] and as a consequence, modify the acute lethal response in mice receiving CY injections. The mode of death from CY was gastrointestinal injury. After CY administration, there was approximately 20% lethality in mice bearing 410 tumors only, 40 to 50% mortality in mice bearing 410 tumors plus a bilateral 168 tumor (or in control mice with no tumors), and 90% in mice bearing the line 168 tumors. Therefore, mice with line 410 neoplasms showed *less* toxicity, while mice bearing 168 neoplasms showed *greater* toxicity than expected. It is possible that such differences arose from tumor influences on hepatic activity. While it is difficult to envision how the presence of a malignancy could modify enzyme activity in the bidirectional manner indicated by the sensitizing and protective effects of 168 and 410 tumors, respectively, such effects might be produced through modification of drug uptake. A less likely possibility might be via effects on *repair* of CY alkylation damage produced in the gut (e.g., the stem cells in the crypts of Lieberkuhn). Clearly, further work is needed in this area, but this situation in which *both* neoplastic and normal tissue sensitivities are affected as a consequence of intratumor heterogeneity is intriguing.

In a study similar to that of Miller et al.,[11,24] Brodt et al.[27] have studied clonal heterogeneity in two spontaneously metastasizing mammary carcinomas in mice (T-58 and MT-2). Three categories of clones were described: (1) rapidly growing clones in vivo, (2) slow-growing clones in vivo, and (3) non-growing clones. When clonal interactions were studied by injection of different clones at various sites, it was found that slow-growing lines failed to modify the growth rates of rapidly growing clones, but would accelerate the growth of a second slow-growing line injected bilaterally. Brodt et al. postulated that clonal interactions may occur only between a restricted number of the total cell subpopulations that may exist in a heterogeneous neoplasm.

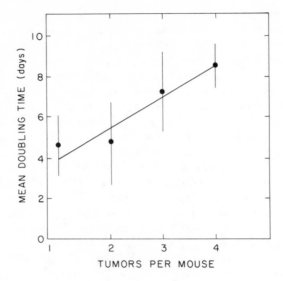

FIGURE 2. The average doubling time for spontaneous mammary tumors in mice as a function of the number of tumors per mouse (error bars are the standard deviations of the mean values).

Chesire[28] has assayed tumor growth rates in mice bearing either one or several spontaneous mammary carcinomas to determine if the number of malignancies present affected overall growth rates. The time taken to double in volume was used as the criterion of potential interaction. When the average doubling times of the spontaneous tumors observed were calculated as function of the total number of neoplasms per mouse, a clear *increase* was noted. This effect is shown in Figure 2. This increase in doubling time appeared linearly related to the number of tumors, and therefore to the total cancer cell burden. The authors felt that this might have been an effect of cachexia.

IV. IN VITRO EXPERIMENTAL STUDIES

Heppner et al.[12] and Miller and colleagues[24] have conducted other studies in vitro, to study clonal interactions among subpopulations of the mouse mammary adenocarcinoma system.

Experiments were performed by placing cover slips in Petri dishes. Cell suspensions were then added, and after the cells attached, the cover slips were transferred to new dishes. Two cover slips bearing either the same or a different tumor subline were used (Figure 3), and these co-cultured lines were exposed to varying concentrations of methotrexate (MTX), a drug that is cell-cycle specific, killing cells in S phase.[29] Cells were treated for 2 to 3 days with MTX, and the number of cells remaining on the cover slips was determined. Tumor cells of lines 410 and 67 were studied, and the concentration of MTX needed to inhibit cell growth by 50% as compared to growth in control cells was determined (ID_{50}) after a treatment time of 72 hrs.

It was found that the average ID_{50} in line 67 cells was 22.7 nM. When line 67 cells were grown in the presence of 410 tumor cells, the ID_{50} was reduced to 10.7 nM. Therefore, the sensitivity of line 67 cells to MTX was increased twofold when cocultured with line 410 cells. A similar increase in sensitivity of line 168 cells was noted when co-cultured with line 410 cells. The authors also found that *no* other clonal sub-

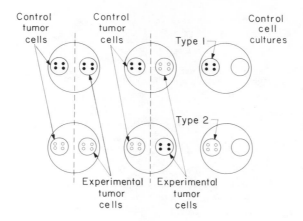

FIGURE 3. Illustration of protocol used with heterogeneous tumor cell subpopulations to investigate clonal interactions in vitro.

populations affected the sensitivity of line 410 cells to MTX, indicating a hierarchical situation for these clonal interactions.

As an ID_{50} assay measures primarily inhibition of growth (i.e., cytostatic activity) and not necessarily actual cell killing (i.e., cytotoxic activity), one should perform some type of clonogenic assay in parallel with ID_{50} studies to factor out these features of clonal interactions. In the in vitro work reported by Miller et al.,[24] it was certainly reasonable to perform ID_{50} studies, as the nature of the expected clonal interactions was growth inhibitory. Still, as cells were exposed to MTX for 48 to 72 hr, which is much longer than the duration of the S phase, some degree of cell killing would also be involved, particularly at high doses of MTX.

In terms of the mechanism of action for the in vitro effects seen with MTX in co-cultured cells, Miller et al.[24] felt that alteration of MTX transport was not the answer. They mention, as other possible rationales, changes in the intracellular levels of dihydrofolate reductase, or modification of endogenous pool sizes of purines, pyrimidines, or folate. Experiments involving the use of conditioned medium from 410 cells (either MTX treated or nontreated) failed to produce any growth inhibition, indicating that if a diffusible factor were produced by 410 cells into the co-cultured medium to affect the target line 67 cells, it is extremely labile.

These data are significant from the viewpoint of intratumor modification of intrinsic drug sensitivity. For example, if a "controlling" tumor line were effectively killed by a particular mode of therapy, this might allow the overt expression of subpopulations that now exhibit an increased "resistance" to some chemotherapeutic agents. However, it is apparent that such vectorial interactions are not well defined, and that appropriate model systems to critically explore such phenomena are of great importance.

Heppner and colleagues have also used the expression of mouse mammary tumor virus antigen (M_uMTV) to study clonal interactions,[2,25] As this mouse mammary carcinoma arose in an animal carrying M_uMTV virus, certain tumor subpopulations (line 68H) strongly express this antigen. Given this immunological diversity, M_uMTV expression could be used to identify subpopulations. In these experiments, line 68H cells were admixed with line 168 cells. When these cells were cultured separately, the cell culture doubling time of 68H cells was about twice that of line 168 cells. In co-cultures, the proportion of cells expressing M_uMTV was monitored as a function of time and it was found that the proportion of 68H cells did *not* change as a function of time in co-culture, although based on the differences in growth rates of the two sub-

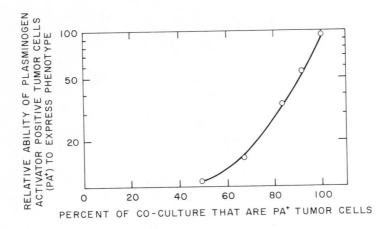

FIGURE 4. Ability of tumor cells to express plasminogen activator (PA) type substances when admixed with a clonal variant that is not able to express PA.

populations, line 68H cells should have exhibited a relative decrease in their percentage composition of the admixed culture. Also, the observed cell culture doubling time was *either* that of 68H alone or of 168 alone, i.e., no intermediate growth rates were noted,[12] showing that while 68H and 168 tumor cells grown independently exhibit large differences in growth rate, growth rates of the two tumor lines when grown together exhibited a strong interaction.

Kyner et al.[30] and Newcomb et al.[31] have published an interesting set of papers dealing with the expression of plasminogen activator (PA) type substances by tumor cells and the relationship of this expression to co-cultivation with nonmalignant cells. Plasminogen activators are increased in cells that are tumorigenic and may be associated with the process of tumor metastasis. Kyner et al. used two mouse melanoma clones (B_559 — a highly tumorigenic clone which produced PA, and C_3471 — a clone which is nontumorigenic and does not produce PA) in co-culture, and investigated the production of PA by line B_559. As can be seen in Figure 4, when the ability of the tumorigenic line to express PA is shown as a function of the total admixture, this ability is strongly quenched by the nontumorigenic C_3471 line. Indeed, a 50% reduction in the expression of PA is produced when the total populations contains roughly only about 10% of the nontumorigenic C_3471 line. The authors were able to exclude such factors as cell cycle perturbation, cell fusion, formation of soluble inhibitors of PA, or presence of other inhibitory substances as being responsible for the observed PA suppression. They felt that close proximity between PA^+ and PA^- cells was required for the effect, although an absolute requirement for cell-to-cell contact was not shown. Such effects might correlate with the presence of unstable inhibitory substances effective only over a short range as suggested by Miller et al.[24] Regardless, the demonstration of inhibition of such a physiological property as expression of PA suggests that the admixtures of phenotypes that occur in a solid tumor could show similar effects. In this regard, Newcomb et al.[31] have studied the in vivo aspects of PA expression and tumorigenicity. B_559 cells routinely produce neoplasms in all mice, while C_3471 cells produce no tumors. When these two lines were admixed in equal proportions, the observed yield was only 10% (rather than 50%), indicating a significant reduction in the tumorigenicity of the B_559 cells. In addition, the latent period was signficantly increased, and PA expression in tumors that did form was decreased by about 75%.

Newcomb et al.[31] also performed experiments in which mice were (1) injected at a single site with admixed B_559 and C_3471 cancer cells, (2) injected at two different sites with either cell line, or (3) injected at the same site, but with a sequential rather than simultaneous set of injections. As expected, tumors appeared in 100% of mice injected with the B_559 cells alone; no tumors were found in mice given C_3471 cells alone. In sites injected with admixed cells, no neoplasms were found, indicating the inhibitory effect of the C_3471 cells. However, in mice receiving the two tumor lines on bilateral sides of the animal, B_559 tumor expression was identical to that seen in control animals, i.e., no inhibition was found. Further, in the mice receiving sequential injections, essentially 100% expression of tumorigenicity was seen (although the latent period was increased by about 55%). These data appear to indicate that close physical contact was necessary between the two cancer lines to produce the inhibition of PA expression. In control experiments in which mice were challenged with immunologically incompatible malignant cell types, no effects were found on the tumorigenicity of B_559 cells, suggesting that inflammatory responses per se were not the cause of the inhibition noted. Newcomb et al.[31] felt that the C_3471 cells suppress the capacity of the B_559 cells to produce PA, which changes the capacity of the B_559 cells to form tumors, possibly through alteration of cell migratory mechanisms.[32] The lack of PA expression (and resultant plasmin formation) may at a later time allow the infiltration of tumoricidal cells such as macrophages.[33]

A different type of clonal interaction clearly involving cell cycle changes has been noted by Dewey et al.,[34] who studied whether cells in the DNA synthetic phase of the cell cycle (S phase) could modify the kinetics of cells in the G_1 phase of the cell cycle. To do this, Chinese hamster ovary (CHO) cells were used, and synchronized mitotic cells were obtained from exponentially growing cultures. These synchronized cells were then either pulse- or continuously labeled with tritiated thymidine (which is incorporated by cells in S phase) and these radioactively labeled cells were identified by autoradiography. In some experiments, the authors co-cultured G_1 and S phase cells together, and in other experiments the investigators used special dishes with recessed lids in which mitotic cells had been placed into the bottom of the dishes. They also plated cells on the bottom of the recessed surfaces of the lids, and by differing the relative times when the bottom surface of the lids or the bottoms of the dishes themselves were plated, they obtained populations of CHO cells in very close apposition (separated by a distance of about 2 mm), but in two different and well-synchronized portions of the cell cycle (i.e., G_1 and S). Then, by following the kinetics of cell progression, the time needed for cells to traverse through the G_1 phase into S phase could be determined. It was found that both conditioned medium obtained from growing of S phase cells, and a close proximity of S phase cells to G_1 phase cells (in co-cultured cells) stimulated G_1 cells to enter the S phase (e.g., a positive interaction). However, this stimulatory effect of S phase cells was lost when the cells were separated by the 2-mm distance. In fact, the authors estimate the maximum effective distance to be about 0.1 mm, because microscopic observations indicated that only S phase cells either touching or within about 0.015-mm distance from a G_1 cell appeared to have a significant effect in inducing G_1 cells to enter S phase. The authors felt that this indicated the existence of two separate inducer substances released from S phase cells. Although CHO cells are not malignant cells, it is not unrealistic to suggest that such intracell cycle interactions could also exist in growing solid tumors. However, no studies involving such quantitative evaluation of potential clonal interactions using cell kinetic approaches have yet been published.

V. IN VIVO METASTATIC ASSAY SYSTEMS

Much work has been done using specific models to study the biology of tumor metastasis and to describe and define aspects of intraneoplastic diversity and clonal interactions.

Poste et al.[35] have performed extensive studies using the B16 mouse melanoma system. This original tumor line has been extensively cloned to isolate variants of different metastatic capability. A variant exhibiting a *high* metastatic potential for lung tissue is the B16-F10 clone, whereas in contrast, a clone termed B16-F1 exhibits a *low* propensity for production of lung metastases.[36,37] These sublines appear to be stable both in vitro and in vivo in terms of their ability to metastasize. For the B16-F1 line, the typical number of visible lung colonies after i.v. injection of 2.5×10^4 cells would be about 13, while in contrast, the B16-F10 line would produce about 117 colonies for the same number of injected cells, indicating that the relative metastatic efficiency of the two tumor sublines is about 9.[35]

However, as pointed out by Poste et al.,[35] the F1 and F10 lines are themselves heterogeneous (albeit stable in terms of metastatic phenotype) and contain subpopulations of cells with differing phenotypic characteristics.[37-39] It is important to note, however, that the *range* of variability in the F10 and F1 lines remains constant with continued passaging of cells, both in vivo and in vitro. This apparent stability is an indirect argument that there exist clonal interactions that serve to stabilize, within certain biological limits, phenotypic properties of tumor cells.

This stability property has consequently been the focus of considerable study. A number of subpopulations were cloned from the F10 (high metastatic potential) parent line (subclones 5, 22, 18, and 42). The number 5 and 22 sublines exhibited a *low* metastatic capacity immediately after subcloning, whereas the number 18 and 42 sublines showed a *high* metastatic capacity. However, this initial level of metastatic potential is *unstable*, as multiple subcultivations in vitro or in vivo yield subclones with significantly different metastatic properties.[35] Comparatively speaking, a greater potential for variation in metastatic phenotype was noted in clones 5 and 22 (originally of low metastatic potential) than in clones 18 or 42 (original high metastatic potential). Poste et al.[35] postulated that this instability in metastatic phenotype argued that the stability noted in the B16-F10 line had to be a result of some type of cooperation among tumor subpopulations. In essence, this cooperation would preserve the expression of the full range of the metastatic phenotype, with the result that such an interactive process would not allow a single (or a few) subpopulation(s) to dominate the response of the B16-F10 tumor line.

To directly investigate this putative stability induced by cooperative events among subpopulations, the authors designed a very clever experiment.[35] Again, the B16-F10 line was cloned, but the objective in this case was to obtain specific drug resistant clones (i.e., to obtain tumor sublines with identifiable phenotypic properties); three specific resistant variants were selected. These variants were resistant either to ouabain, trifluorothymidine, or diaminopurine, and were denoted as Oua^r, TFT^r, or DAP^r cells, respectively. These resistant variants could then be doubly selected to have degrees of metastatic capability from low to high. Therefore, Poste et al.[35] could perform experiments in which wild type (i.e., drug sensitive) cells were mixed with drug resistant lines, all possessing any desired degree of metastatic potential. As a consequence of having two phenotypic markers (i.e., drug resistance and metastatic capability), changes in the numbers of lung metastases observed could be related to the initial composition of the population. These drug-resistant clones were stable when grown as a polyclonal culture.

To gain additional insight into the dynamics of this stability, Poste et al.[35] purposefully induced instability into their experimental system. Specifically, they began with a cell admixture consisting of TFT^r cells (which produce about 20 lung colonies per 2.5

× 10⁴ cells), Oua^r cells (which produce about 120 lung colonies/2.5 × 10⁴ cells), and a double resistant TFT^r/Oua^r line (which produces about 200 lung colonies/2.5 × 10⁴ cells). Therefore, the metastatic potentials of these resistant lines were clearly different, and this three-component system was stable. Instability was introduced by treating the admixed cultures with TFT, and surviving cells were assayed for metastatic ability. As expected, the Oua^r cells were selectively killed. However, with continued cultivation of the (now) two-component system (i.e., TFT^r and TFT^r/Oua^r cells), it was found that instead of just exhibiting low (i.e., about 20 lung colonies) and high (i.e., about 200 lung colonies) metastatic abilities, that by 10 subcultivations in vitro, variants with *intermediate* metastatic capability had arisen in the population, i.e., there was reestablishment of the range of the metastatic phenotype following introduction of instability. Both the low and high metastatic component appeared to contribute to this reestablishment. By 20 subcultivations, a new stable situation had become established. This new system then had DAP^r cells added to it (these cells produce about 50 lung colonies/2.5 × 10⁴ cells), the system was allowed to stabilize, and was then treated with DAP. As expected, in the initial passages, only DAP^r survivors were found; but again, by 20 subcultivations, a wide range of metastatic potential was seen (by a factor of about 10 from lowest to highest metastatic potential). This illustrates that restriction of the number of subpopulations apparently introduces instability in terms of metastatic potential into the remaining subpopulation(s).

If this situation represents a general phenomenon, it would appear to be very important. One could envision that any external selection pressure (i.e., cytotoxic drugs, ionizing radiation, hyperthermia, etc.) that would serve to differentially affect tumor subpopulations would necessarily affect intratumor stability. As a consequence, then, the phenotypic expression of the remaining subpopulations may change. While the previous discussion has centered around metastatic behavior as the end point (and it is clearly an important, clinically relevant end point), it is important to note that these investigators did not feel that other phenotypic properties may likewise exhibit such instability. Their work with drug resistant clones indicated that this property (drug resistance) remains stable even in the face of the metastatic instability. Also, it would appear that the generation of metastatic instability in surviving subpopulations of a heterogeneous tumor after exposure to a selection pressure may be *independent* of the agent used to generate the instability, i.e., the process of population restriction and subsequent (metastatic) diversification will occur after *any* type of selection pressure. This process must be intimately related to the cell kinetics and proliferative behavior of the cells comprising the tumor and is obviously a complex situation (c.f., Chapter 5). Further, it has been shown by examination of metastatic lung colonies from mice sacrificed at different times postinjection of tumor cells (e.g., 18 vs. 40 days) that by 40 days the lung colonies were themselves heterogeneous in terms of their metastatic capabilities (intralesional heterogeneity).[40]

This phenomenon of intratumor (metastatic) stability has also been reported in other murine systems, the UV2237 fibrosarcoma,[41] the RAW117 lymphosarcoma,[42] the rat 13762 mammary adenocarcinoma[43,44] and the rat 1AR6 hepatocarcinoma.[45]

Poste et al.[46] have recently postulated that tumors early in their evolution may be populated by clones that have stable metastatic properties, and that as the tumor progresses there is the emergence of clonal subpopulations that possess unstable metastatic phenotypes. More specifically, they postulate an initial phase (Phase 1) in which there exist clones with stable, but diverse phenotypic properties, and that these populations expand with time. Later, there is a second phase (Phase 2) in which there is evolution of subpopulations with unstable metastatic properties, and that these clones "are stabilized by interactions with other clones."[46] Finally, there is a third phase (Phase 3)

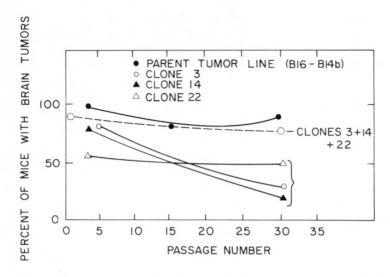

FIGURE 5. Effects of clonal stability on ability to produce metastatic brain tumors in mice after i.v. injection. Tumor clones (clones 3, 14, and 22) were selected from a B16-B14 b parent melamona tumor line. When clones were injected separately, they produced a low level of brain tumors; however, when they were admixed, a high level of brain tumors were seen, indicating the presence of clonal interactions.

characterized by the emergence of subpopulations that are unstable with regard to the metastatic phenotype, but are now "unresponsive to the stabilizing effects of other clones."[46]

Other work using the B16 mouse melanoma system has been reported by Miner et al.[47] They studied the phenotypic stability of a stable brain-colonizing B16 subline termed B16-B14b. This variant was selected 14 times from B16-F1 (low metastatic potential),[36] and typically i.v. injection of 2×10^4 cells would yield brain tumors in all mice. These neoplasms were morphologically heterogeneous, as both pigmented and nonpigmented tumor colonies could be seen. To investigate possible clonal interactions, 18 different subclones from the parental B16-B14b line were obtained which were markedly heterogeneous in morphology, melanin content, and ability to produce brain tumor colonies. However, it was found that a relative stabilization of brain-colonizing potential could be achieved by admixing (in equal proportions) and co-culturing some of the different clones. This result is shown in Figure 5, in which it may be seen that, although the colonizing potential of clones 3, 14, or 22 decreased significantly with time in tissue culture, the co-cultured admixture of these clones still yielded a brain-colonizing potential equivalent to that of the parental B16-B14b tumor line for up to 31 passages in vitro. That this effect was dependent on mutual growth interactions was demonstrated by simply mixing the three clones together immediately before injection: this did not produce an effect different from the clones injected separately (i.e., no stabilization effect was seen). The authors also note that, whereas brain-colonizing potential could be stabilized by polyclonal growth, no such relationship existed for other phenotypic properties, such as cell growth rate, melanin production, or morphology.

Other studies have been focused on the relationship of the growth of a primary tumor to the growth of metastases in the same host. For example, DeWys[48] studied the growth kinetics of the nonimmunogenic Lewis lung carcinoma transplanted into the leg muscles of mice. To follow the growth of metastases, the primary neoplasm was amputated. Metastatic growth in the lungs could be detected by 9 days later, and kidney metastases by 17 days after implantation of the primary tumor burden. DeWys[48] found that tumor growth in the different sites (leg, lung, kidney) was comparable in all locations, particularly in the late stages of growth. When a second tumor was implanted in the contralateral leg 8 to 15 days after the initial tumor implant, however, it was found that the growth rates of these second neoplasms slowed much more abruptly and reached a plateau at a much smaller size. This was interpreted as indicating the operation of systemic factors which were dependent on the presence (and size of) an initial tumor. Therefore, the implication was that large primary malignancies *inhibit* the growth of secondary tumors. This conclusion was supported by the observations that removal of the primary tumor (by amputation) would partially reverse some of this growth inhibiting effect, and that the reversal of this inhibition decreased the longer the primary cancer was present. Also, the *total* cancer cell burden (i.e., the sum of the weight of primary and secondary neoplasms) as a function of time seemed to be the same as for primary tumors growing alone,[48] suggesting that systemic inhibitory factors were directly related to the total tumor burden. DeWys[48] also noted that similar effects have been reported by Goodman,[49] who felt that such effects represented interference with the general nutrition of the host, although release of cytostatic or cytotoxic by-products from (necrotic) tumor tissue was also possible.

Greene and Harvey[50] have performed similar experiments to DeWys,[48] using a transplantable hamster lymphoma. In this system, metastases do not normally develop. However, the investigators showed that this lack of metastases was *not* due to the fact that there were no circulating cancer cells, as blood taken from hamsters with lymphoma and reinjected into control animals did produce tumors. In contrast, blood from lymphoma-bearing animals reintroduced into lymphoma-bearing hamsters did *not* produce tumors, and Greene and Harvey concluded that the presence of the primary neoplasm "exerted an inhibitory influence on the growth of its cells transplanted either manually or by circulating blood elsewhere in the animal's body."[50] When the primary neoplasm was removed from the animals, about two thirds of the hamsters showed metastatic disease within 3 to 4 weeks. To prove that surgical manipulation per se was not the reason from this increase in metastatic appearance, experiments were performed in which only partial tumor removal was done. Results were consistent with the fact that manipulation with subsequent seeding of cancer cells into the lymphatic and blood vascular systems was not the cause of the increase in metastatic frequency seen. The authors hypothesized that the lack of development of metastases was due to inhibitory effects on vascularization of metastases rather than against neoplastic cell growth.

Price et al.[51] have also done experiments relevant to the interactions among multiple neoplasms in the same host. Using spontaneously rising mouse mammary cancers, the investigators studied whether separately arising tumors in the same mouse had similar or different lung colonization potential (LCP). In mice showing multiple tumors, about half had tumors showing LCP of a low degree; about 10% had tumors of similar LCP of moderate degree; about 10 to 20% showed tumors of similar LCP of high degree; and about 30% showed tumors of *disparate* (.e., both high and low) LCP. Therefore, neoplasms of varying degree of metastatic potential can arise in the same host. There is no indication from this data that lung metastatic potential is modulated by clonal interactions between individual tumors; if there were such interactions, one might postulate that multiple neoplasms arising within a given host should demonstrate very similar metastatic potentials.

Miller[15] and Miller et al.[16,52] have used the mouse mammary adenocarcinoma system developed by Heppner and co-workers[10,18] to investigate metastatic aspects of clonal interactions, as these mammary tumor lines differ significantly in their metastatic be-

Table 1
INCIDENCE OF METASTASIS (PERCENT) FROM MOUSE MAMMARY ADENOCARCINOMA SUBPOPULATIONS AS A FUNCTION OF TUMOR SITE

Tumor subpopulation	Percent metastases	
	Subcutis	Fatpad
66	85	90
67	0	6
168	3	3
68H	0	0
410.4[a]	82	92
410	8	0
410a	45	67
4501	66	71
4526	83	89

[a] 410.4 cells are tissue culture-adapted tumor cells from the original 410LM metastatic tumor; 410 cells are a low metastatic variant of 410LM cells; 410a is a high metastatic subclone of the 410 tumor line; 4501 and 4526 are subclones that are in vitro adapted subclones derived from an in vivo 410.4 tumor.

Data taken from Miller, F. R., Miller, B. E., and Heppner, G. H., *Invasion Metastasis*, 3, 22, 1983.

Table 2
APPROXIMATE NUMBERS OF TUMOR CELLS FROM A HETEROGENEOUS MOUSE MAMMARY ADENOCARCINOMA TUMOR SYSTEM NEEDED TO PRODUCE LUNG METASTASES IN ALL RECIPIENTS

Tumor subpopulation	Approximate tumor cell number
66	1×10^5
67	1×10^7
168	1×10^6
68H	$> 10^7$
410.4	3×10^4
410	5×10^6
410a	(no data)
4501	1×10^6
4526	1×10^6

[a] Data taken from Miller, F. R., Miller, B. E., and Heppner, G. H., *Invasion Metastasis*, 3, 22, 1983. With permission.

havior. In Table 1, the percentages of mice with metastases after injection of tumor cells in two different sites in the animal are shown.[15,16] Of the original lines developed from the primary tumor, line 66 had the highest metastatic propensity. A number of variants of the original metastatic lung tumor line (410 LM) were also studied, and with the exception of line 410, also appeared to have a high metastatic potential. In Table 2, the approximate numbers of tumor cells needed to produce lung metastases in mice are listed, and as expected, lines 66 and 410 were the most proficient in their metastatic capabilities.[52] Knowing the independent metastatic behavior of these tumor sublines, Miller investigated some possible tumor subpopulation interactions. In an initial experiment, mice were injected i.v. with either line 67 cells or line 410.4 cells, or with an admixture of the two cell lines. Metastatic lung colonies were assayed 5 weeks postinjection. It was found that the line 67 cells by themselves produced no lung colonies, while the line 410.4 cells produced lung colonies in only 1 of 5 mice. With the admixture of cells, lung nodules were found in 10 of 11 mice. Therefore, Miller felt that clonal interactions were present. Individual metastases were taken from the mice receiving the admixture of lines 67 and 410.4 tumor cells and cultured in vitro. Miller visually identified the outgrowing cell colonies by cell morphology and found that of 13 metastatic nodules investigated, 9 appeared to contain purely 410.4 cells, 1 contained only line 67 cells, and the remaining 3 contained both 67 and 410.4 cells. Miller therefore hypothesized that line 67 could increase the metastatic efficiency of line 410.4 cells.[15]

Miller[15] also injected mice subcutaneously with line 168 tumor cells, and then subsequently reinjected the mice intravenously with line 168 cells, or with line 410.4 cells. Line 168 cells as subcutaneous solid tumors show a low incidence of production of metastatic lung tumors,[16] whereas 410.4 tumor cells injected i.v. are efficient in forming lung nodules. No mice in which line 168 solid tumors were grown alone, or in animals with *in situ* line 168 tumors which additionally were injected i.v., with line 168 cells, showed evidence of lung metastases, while about one half of the mice given 410.4 cells showed lung metastases. When individual metastatic nodules were cultured in vitro, again using morphology as a basis of subpopulation identification, all metastases contained 410.4 cells. However, 2 of 9 metastases were heterogeneous in that they also contained 168 cells. This finding may indicate that the establishment in the lung of a metastatic tumor subpopulation (i.e., line 410.4) can act to promote the metastasis of normally nonmetastatic tumor lines, possibly through an increased trapping of cells which probably circulate in the bloodstream, but do not by themselves form lung colonies. This could point to a mechanism for the generation of heterogeneous secondary neoplasms that does not depend on the phenomenon of progression for development of intratumor heterogeneity in metastatic sites.[53,54]

VI. TUMOR-HOST CELL INTERACTIONS

The previous discussion shows that there are multiple mechanisms through which neoplastic cell subpopulations may interact and modulate phenotypic expression. it should also not be forgotten that malignancies contain different proportions of neoplastic and normal tissue cells (cf., Chapter 5).[55-58] As the percentage of normal tissue cells may indeed be quite high (i.e., 30 to 60% of the entire tumor),[59] there is obviously a potentially major role for possible modification of cancer cell expression by the surrounding normal tissue(s). Of further importance is that these normal tissues are in a state of dynamic flux, and the relative proportions of host cells will change with tumor growth and as a function of any intercedent cytotoxic treatment (cf., Chapter 8). Such heterogeneity in macrophage composition has been described by Heppner et al.[14] using the mouse mammary adenocarcinoma system. In 1978, Blazar and Heppner[60] showed that the major lymphocyte population in these mammary carcinomas was either a T lymphocyte or a null cell, and that B type lymphocytes were seldom encountered, and recently, Rios et al.[61] have established that heterogeneity in these normal tissue elements exists among the solid tumors established from subpopulations of the original adenocarcinoma. Using different antibodies to identify subpopulations of T lymphocytes, Rios et al.[61] showed that the most strongly expressed antigen by T cells in line 168 and 410 tumors was Lyt 2.2, while T cells in line 68H tumors most strongly expressed Thy 1.2 and Lyt 1.2 antigens. Therefore, there is heterogeneity in the T-cell component found in a given neoplasm. Further, Rios et al.[61] showed that when carcinomas of two different types were grown on the contralateral sides of mice (i.e., 68H and 168 tumors), there was *no* modification of the respective antigen expressions of T cells. In terms of the macrophage contents, Heppner et al.[14] state that 66, 410.4, 68H, and 168 carcinomas contained 35, 36, 32, 44, and 20%, respectively. While there were differences in macrophage content (e.g., lines 67 and 168), these differences did not correlate with metastatic potential, growth rate, or immunogenicity. However, Heppner et al.[14] suggested that the macrophages in line 66 and 4.10 tumors were apparently more "mature", in that they were found in a late fraction of an elutriation centrifugation separation of cells from these neoplasms (cf., Chapter 5). Appearances of macrophages in a late fraction in such a cell separation experiment would be due to increased size and density of cells, and this illustrates that not only do the macrophage contents of tumors vary, but that the relative maturity of these cells may also vary. As mature macrophages are more effective in cytotoxic activity, this may represent one aspect of host-tumor interactions. While it is not apparent why such differential macrophage maturity should occur, this situation could easily affect tumor cells. For example, Weitzmann and Stossel speculate that as mature macrophages produce substances that may be mutagenic, this may be a potential mechanism for tumor progression.[62,63] Another illustration of tumor-host cell interactions was shown by Henry et al.[64] who demonstrated in co-cultures of macrophages and clonal subpopulations from the Lewis lung carcinoma that collagen degradation by the carcinoma cells (i.e., "invasiveness") was induced by a soluble factor (a monokine) produced by the macrophages.

VII. SUMMARY

The research discussed in this chapter shows that the phenomenon of "clonal interactions" does occur and that models illustrating this have been developed. However, it is also readily apparent that there is little data to explicitly document specific mechanisms through which such clonal interactions are expressed. An attempt to mathematically model tumor heterogeneity and cellular interactions has been made recently by Kendal.[65] Kendal related the volumetric growth of tumors to their degree of heterogeneity by use of methods from statistical mechanics. Using this approach, Kendal showed that interactions between dissimilar cell subpopulations influencing growth yielded the well-known Gompertz equation.[66] Therefore, it seems that the Gompertz equation could be used to describe intratumor heterogeneity in a more sophisticated manner than has been done in the past. Efforts such as this are needed.

In summary, interactions between cellular subpopulations have been shown for:

1. In vivo studies of unperturbed neoplasms in which endpoints such as growth rates have been studied in situation where multiple tumors derived from a common ancestor have been present in the same host
2. In vivo studies of perturbed neoplasms in which the differential responses to cytotoxic agents have been studied
3. In vivo studies in which the process of metastasis has been used as the endpoint to study aspect of tumor subpopulation stability and the relationship of primary tumors to their metastases
4. In vitro studies in which tumor subpopulations have been grown together (co-cultured) and aspects of cell survival or cell kinetics have been assayed

Where interactions have been noted, these effects have been both positive and negative in direction, and it is obvious that a wide variety of mechanisms may be operative. It would seem that one simple way to distinguish such phenomena could be on the basis of proximity, i.e., "near" interactions vs. "far" interactions.

For example, "near" interactions could then include responses dependent on cell-to-cell communication. Such interactions would likely then be mediated through membrane apposition probably with formation of gap junctions, which would allow intercellular passage of small molecular weight compounds. Such mechanisms would likely then be positive or "protective" in the nature of their actions. Other "near" but not necessarily intimate mechanisms could be due to release of biochemical products which could be active (in either a positive or a negative sense) in the local vicinity of other cells. Such humoral or "paracrine" factors would act over a relatively short distance, and this effective distance would be vitally dependent on the stability (lability) of the agent. Clearly, a second important factor in the response of the target cell(s) would be mechanisms for recognition (e.g., receptors) and transport to internal sites (e.g., the DNA).

By "far" interactions one would clearly have to postulate the presence of relatively stable means of cell interaction. These could include stable biochemical products produced by tumor subpopulations, as well as immunological responses of the host, and would most easily explain some of the systemic in vivo effects described in this chapter.

Our discussion of "near" vs. "far" interactions would relate directly to the situations of interactions within a neoplasm vs. the interactions between a primary and a metastatic tumor or between different metastases.

It is important to note at this point that very little is known about the existence of such clonal interactions in human neoplasms. The relevance of the findings in the rodent models clearly needs to be extended to the situation of heterogeneous human cancers. As the natural history of human tumors is different from that of rodent tumors (cf., Chapter 2), particularly with regard to the overall time course of neoplasia, it may be that the "potency" of clonal interactions in human tumors is also different, and it is quite possible that they may not even be demonstrable. This uncertainty again illustrates that quantitative assays of sufficient resolving power to clearly prove the presence or absence of such clonal interactions are absolutely critical.

REFERENCES

1. Heppner, G. H., Cited in tumors: a mixed bag of cells, *Science,* 215, 275, 1982.
2. Heppner, G. H., Tumor subpopulation interactions, in *Tumor Cell Heterogeneity: Origins and Implications,* Owens, A. H., Jr., Coffey, D. S., and Baylin, S. B., Eds., Academic Press, New York, 1982, 225.
3. Woodruff, M. F. A., Cellular heterogeneity in tumours, *Br. J. Cancer,* 47, 589, 1983.
4. Miller, B. E., Miller, F. R., and Heppner, G. H., Assessing tumor drug sensitivies by a new *in vitro* assay which preserves tumor heterogeneity and subpopulation interactions, *J. Cell Physiol.,* Suppl. 3, 105, 1984.
5. Nicolson, G. L., Cell surface molecules and tumor metastasis. Regulation of metastatic phenotypic diversity, *Exp. Cell Res.,* 150, 3, 1984.
6. Heppner, G. H., Miller, B. E., and Miller, F. R., Tumor subpopulations in neoplasma, *Biochim. Biophys. Acta,* 695, 215, 1983.
7. Heppner, G. H. and Miller, B. E., Tumor heterogeneity: biological implications and therapeutic consequences, *Cancer Metastasis Rev.,* 2, 5, 1983.
8. Schirrmacher, V., Shifts in tumor cell phenotypes induced by signals from the microenvironment: relevance for the immunobiology of cancer metastasis, *Immunobiology,* 157, 89, 1980.
9. Poste, G. and Fidler, I. J., Cited in tumors: a mixed bag of cells, *Science,* 215, 275, 1982.
10. Heppner, G. H., Dexter, D. L., DeNucci, T., Miller, F. R., and Calabresi, P., Heterogeneity in drug sensitivity among tumor cell subpopulations of single mammary tumor, *Cancer Res.,* 38, 3758, 1978.
11. Miller, B. E., Miller, F. R., Leith, J., and Heppner, G. H., Growth interaction *in vivo* between tumor subpopulations derived from a single mouse mammary tumor, *Cancer Res.,* 40, 3977, 1980.
12. Heppner, G., Miller, B., Cooper, D. N., and Miller, F. R., Growth interactions between mammary tumor cells, in *Cell Biology of Breast Cancer,* McGrath, C. M., Brennan, M. J., and Rich, M. A., Eds., New York, Academic Press, 1980, 161.
13. Heppner, G. H., Shapiro, W. R., and Rankin, J. C., Tumor heterogeneity, in *Pediatric Oncology,* Vol. 1, Humphrey, G. B., Ed., Martinus Nijhoff, The Hague, 1980.
14. Heppner, G. H., Loveless, S. E., Miller, F. R., Mahoney, K. H., and Fulton, A. M., Mammary tumor heterogeneity, in *Cancer Invasion and Metastasis: Biologic and Therapeutic Aspects,* Nicolson, G. L. and Milas, L., Eds., Raven Press, New York, 1984, 209.
15. Miller, F. R., Tumor subpopulation interactions in metastasis, *Invasion Metastasis,* 3, 234, 1983.
16. Miller, F. R., Miller, B. E., and Heppner, G. H., Characterization of metastatic heterogeneity among subpopulations of a single mouse mammary tumor: heterogeneity in phenotypic stability, *Invasion Metastasis,* 3, 22, 1983.
17. Hager, J. C., Russo, R. L., Ceriani, J. A., Peterson, S., Fligiel, Z., Jolly, G., and Heppner, G. H., Epithelial characteristics of five subpopulations of a heterogeneous strain BALB/cfC$_3$H mouse mammary tumor, *Cancer Res.,* 41, 1720, 1981.

18. Dexter, D. L., Kowalski, H. M., Blazar, B. A., Fligiel, Z., Vogel, R., and Heppner, G. H., Heterogeneity of tumor cells from a single mouse mammary tumor, *Cancer Res.,* 38, 3174, 1978.
19. Calabresi, P., Dexter, D. L., and Heppner, G. H., Clinical and pharmacological implications of cancer cell differentiation and heterogeneity, *Biochem. Pharmacol.,* 28, 1933, 1979.
20. DeWyngaert, J. K., Leith, J. T., Peck, R. A., Jr., Bliven, S. F., Zeman, E. M., Marino, S. A., and Glicksman, A. S., Differential RBE values obtained for mammary adenocarcinoma tumor cell subpopulations after 14.8 MeV neutron irradiation, *Radiat. Res.,* 88, 118, 1981.
21. Brenner, H. J., Leith, J. T., DeWyngaert, J. K., Dexter, D. L., Calabresi, P., and Glicksman, A. S., Protection against hyperthermic cell killing of mouse mammary adenocarcinoma cells *in vitro* by N,N-dimethylformamide, *Radiat. Res.,* 88, 291, 1981.
22. Galli, S. J., Bast, R. C., Jr., Bast, B. S., Isomura, T., Zbar, B., Rapp, H. J., and Dvorak, H. F., Bystander suppression of tumor growth: evidence that specific targets and bystanders are damaged by injury to a common microvasculature, *J. Immunol.,* 129, 890, 1982.
23. Nagai, A., Zbar, B., Terata, N., and Hovis, J., Rejection of retrovirus-infected tumor cells in guinea pigs: effect on bystander tumor cells, *Cancer Res.,* 43, 5783, 1983.
24. Miller, B. E., Miller, F. R., and Heppner, G. H., Interactions between tumor subpopulations affecting their sensitivity to the antineoplastic agents cyclophosphamide and methotrexate, *Cancer Res.,* 41, 4378, 1981.
25. Miller, F. R. and Heppner, G. H., Immunologic heterogeneity of tumor cell subpopulations from a single mouse mammary tumor, *J. Natl. Cancer Inst.,* 63, 1457, 1979.
26. Calabresi, P. and Parks, R. E., Jr., Alkylating agents, antimetabolites, hormones, and other antiproliferative agents, in *The Pharmacological Basis of Therapeutics,* 5th ed., Goodman, L. S. and Gilman, A., Eds., Macmillan Publ., New York, 1975, 1254.
27. Brodt, P., Parhar, R., Sankar, P., and Lala, P. K., Studies on clonal heterogeneity in two spontaneously metastasizing mammary carcinomas of recent origin, *Int. J. Cancer,* 35, 265, 1985.
28. Chesire, P. J., Effect of multiple tumours on mammary tumour growth rates in the C_3H mouse, *Br. J. Cancer,* 542, 1970.
29. Pratt, W. B. and Ruddon, R. W., Eds., The Anticancer Drugs, Oxford Univ. Press, New York, 1979, 98.
30. Kyner, D., Christman, J., Acs, G., Silagi, S., Newcomb, E. W., and Silverstein, S. C., Co-cultivation of tumorigenic mouse melanoma cells with cells of a non-tumorigenic subclone inhibits plasminogen activator expression by the melanoma cells, *J. Cell. Physiol.,* 95, 159, 1978.
31. Newcomb, E. W., Silverstein, S. C., and Silagi, S., Malignant mouse melanoma cells do not form tumors when mixed with cells of a non-malignant subclone: relationships between plasminogen activator expression by the tumor cells and the host's immune response, *J. Cell Physiol.,* 95, 169, 1978.
32. Ossowski, L., Quigley, J. P., and Reich, E., Plasminogen, a necessary factor for cell migration in vitro, in *Proteases and Biological Control,* Cold Spring Harbor Laboratory, New York, 1975, 901.
33. Roblin, R. O., Hammond, M. E., Bensky, N. D., Dvorak, A. M., Dvorak, H. F., and Black, P. H., Generation of macrophage migration inhibitory activity by plasminogen activators, *Proc. Natl. Acad. Sci. U.S.A.,* 74, 1570, 1977.
34. Dewey, W. C., Miller, H. H., and Nagasawa, H., Interactions between S and G_1 cells, *Exp. Cell Res.,* 77, 73, 1973.
35. Poste, G., Doll, J., and Fidler, I. J., Interactions among tumor subpopulations affect stability of the metastatic phenotype in polyclonal populations of B16 melanoma cells, *Proc. Natl. Acad. Sci. U.S.A.,* 78, 6226, 1981.
36. Fidler, I. J., Selection of successive tumor lines for metastasis, *Nature (London) New Biol.,* 242, 148, 1973.
37. Fidler, I. J., Gersten, D. M., and Budmen, M. B., Characterization *in vivo* and *in vitro* of tumor cells selected for resistance to syngeneic lymphocyte-mediated cytotoxicity, *Cancer Res.,* 36, 3160, 1976.
38. Fidler, I. J. and Kripke, M. L., Metastasis results from pre-existing variant cells within a malignant tumor, *Science,* 197, 893, 1977.
39. Hart, I. R., The selection and characterization of an invasive variant of the B16 melanoma, *Am. J. Pathol.,* 97, 587, 1979.
40. Poste, G., Doll, J., Brown, A. E., Tzeng, J., and Zeidman, I., A comparison of the metastatic properties of B16 melanoma clones isolated from cultured all lines, subcutaneous tumors and individual lung tumors, *Cancer Res.,* 42, 2770, 1982.
41. Cifone, M. and Fidler, I. J., Increasing metastatic potential is associated with increasing genetic instability of clones isolated from murine neoplasms, *Proc. Natl. Acad. Sci. U.S.A.,* 78, 6949, 1981.
42. Nicolson, G. L., cited in an article by Fidler, I. J. and Poste, G., The heterogeneity of metastatic properties in malignant tumor cells and regulation of the metastatic phenotype, in *Tumor Cell Heterogeneity: Origins and Implications,* Owens, A. H., Jr., Coffey, D. S., and Baylin, S. B., Eds., Academic Press, New York, 1982, 138.

43. Nicolson, G. L., Cell surface and cancer metastasis, *Hosp. Pract.*, 18, 75, 1982.
44. Nicolson, G. L., An introduction to cancer invasion and metastasis, in *Cancer Invasion and Metastasis: Biologic and Therapeutic Aspects*, Nicolson, G. L. and Milas, L., Eds., Raven Press, New York, 1983, 1.
45. Talmadge, J. E., Starkey, J. R., and Stanford, D. R., In vitro characteristics of metastatic variant subclones of restricted genetic origin, *J. Supramol. Struct. Cell Biochem.*, 15, 139, 1981.
46. Poste, G., Greig, R., Tzeng, J., Koestler, T., and Corwin, S., Interactions between tumor cell subpopulations in malignant tumors, in *Cancer Invasion and Metastasis: Biologic and Therapeutic Aspects*, Nicolson, G. L. and Milas, L., Eds., Raven Press, New York, 1984, 223.
47. Miner, K. M., Kawaguchi, T., Uba, G. W., and Nicolson, G. L., Clonal drift of cell surface, melanogenic, and experimental metastatic properties of *in vivo*-selected, brain meninges — colonizing murine B16 melanoma, *Cancer Res.*, 42, 4631, 1982.
48. DeWys, W. D. Studies correlating the growth rate of a tumor and its metastases and providing evidence for tumor-related growth-retarding factors, *Cancer Res.*, 32, 374, 1972.
49. Goodman, G. J., Effects of one tumor upon the growth of another, *Proc. Am. Assoc. Cancer Res.*, 2, 207, 1957.
50. Greene, H. S. N. and Harvey, E. K., The inhibitory influence of a transplanted hamster lymphoma on metastasis, *Cancer Res.*, 20, 1094, 1960.
51. Price, J. E., Carr, D., Jones, L. D., Messer, P., and Tarin, D., Experimental analysis of factors affecting metastatic spread using naturally occurring tumors, *Invasion Metastasis*, 2, 77, 1982.
52. Miller, F. R., Medina, D., and Heppner, G. H., Preferential growth of mammary tumors in intact mammary fatpads, *Cancer Res.*, 41, 3863, 1981.
53. Nowell, P. C., The clonal evolution of tumor cell populations, *Science*, 194, 23, 1976.
54. Nowell, P. C., Tumors as clonal proliferation, *Virchows Arch. B: (Cell Pathol.)*, 29, 145, 1978.
55. Evans, R., Macrophages in syngeneic animal tumors, *Transplantation*, 14, 468, 1972.
56. Russell, S. W., Doe, W. F., Hoskins, R. G., and Cochrane, D. G., Inflammatory cells in solid murine neoplasms. I. Tumor disaggregation and identification of constituent inflammatory cells, *Int. J. Cancer*, 18, 322, 1976.
57. Lord, E. M. and Keng, P. C., Effects of radiation on *in situ* host cells separated from a murine tumor by centrifugal elutriation, *Radiat. Res.*, 83, 456, 1980.
58. Stewart, C. C. and Beetham, P. L., Cytocidal activity and proliferation ability of macrophages infiltrating the EMT6 tumor, *Int. J. Cancer*, 22, 152, 1978.
59. Siemann, D. W., Lord, E. M., Keng, P. C., and Wheeler, K. T., Cell subpopulations dispersed from solid tumours and separated by centrifugal elutriation, *Br. J. Cancer*, 44, 100, 1981.
60. Blazar, B. A. and Heppner, G. H., *In situ* lymphoid cells of mouse mammary tumors. II. The characterization of lymphoid cells separated from mouse mammary tumors, *J. Immunol.*, 120, 1881, 1978.
61. Rios, A. M., Miller, F. R., and Heppner, G. H., Characterization of tumor-associated lymphocytes in a series of mouse mammary tumor lines with differing biological properties, *Cancer Immunol. Immunother.*, 15, 87, 1983.
62. Weitzmann, S. A. and Stossel, T. P., Mutation caused by human phagocytes, *Science*, 212, 546, 1981.
63. Weitzmann, S. A. and Stossel, T. P., Effects of oxygen radical scavengers and antioxidants on phagocyte-induced mutagenesis, *J. Immunol.*, 128, 2770, 1982.
64. Henry, N., van Lamsweerde, A.-L., and Vaes, G., Collagen degradation by metastatic variants of Lewis lung carcinoma: cooperation between tumor cells and macrophages, *Cancer Res.*, 43, 5321, 1983.
65. Kendal, W. S., Gompertzian growth as a consequence of tumor heterogeneity, *Math. Biosci.*, 73, 103, 1985.
66. Laird, A. K., Dynamics of tumor growth: comparison of growth rates and extrapolation of the growth curve to one cell, *Br. J. Cancer*, 19, 278, 1965.

Chapter 8

RESPONSES OF PRIMARY HETEROGENEOUS TUMORS TO THERAPY

I. INTRODUCTION

In previous chapters we have described in depth various aspects of the differential phenotypic expression of heterogeneous tumors, both animal and human. The purpose of this chapter is to examine another aspect of the neoplastic phenotype, namely, the response to agents that are used, or may be used in the clinical treatment of cancer. The chapter will focus on how we presume the subpopulations of a primary solid tumor would respond to modalities such as chemotherapeutic agents (including differentiation compounds), ionizing radiation, and hyperthermia, either as single agents or in combination. In this discussion, the reader should keep in mind the evidence presented in Chapters 5 to 7 concerning cell kinetic and environmental factors and potential clonal interactions, which will modulate and modify the responses of tumor subpopulations to any given agent. In this chapter, we will focus on subpopulations of a primary neoplasm although some comparison will be made to responses of metastatic subpopulations. The therapy of metastatic disease is further discussed in detail in Chapter 9.

II. RESPONSES TO CHEMOTHERAPEUTIC AGENTS

In the progression of neoplastic disease to advanced cancer, resistance to chemotherapy given on a combination basis is often found. The basis for such resistance is likely due to treatment selection of phenotypically resistant tumor subpopulations. In this regard, it is important to describe how subpopulations of tumor cells respond to chemotherapeutic agents.

Such differential responses of heterogeneous subpopulations have been studied in the human colon adenocarcinoma (DLD-1) system.[1-3] This tumor system has been described previously in this text (Chapter 2), and in vitro drug sensitivity studies have been performed with this system using the ID_{50} endpoint (i.e., the amount of drug required to inhibit cell growth by a factor of 50%); relevant data are presented in Table 1. There is diversity in the effects of these cytotoxic drugs upon the tumor populations. Actinomycin D was the most potent drug used (in terms of the concentration required to produce growth inhibition). Also, whereas the parent line (DLD-1) was the most sensitive to actinomycin D, the two clonal subpopulations (A and D) were of equal sensitivity.

In contrast, MeCCNU (semustine) was the least effective agent investigated. A major difference was noted between the responses of the A and D subpopulations with respect to sensitivity to mitomycin C, with clone A being about 30 times more sensitive than clone D. As mitomycin C produces cross-linking in DNA, it is attractive to postulate that the increased sensitivity of clone A is associated with its aneuploid status (e.g., median chromosome number of 76) as contrasted to the diploid clone D line. Regardless, it is clear that differential sensitivities exist among these three tumor lines with respect to their responsivity to commonly used chemotherapeutic agents, and that the absolute ranking of sensitivity varies depending on the drug studied.[3]

A similar type of study has been conducted by Heppner et al.[4] using the heterogeneous mouse mammary adenocarcinoma system (cf., Chapters 1 and 7). Both in vitro and in vivo studies were performed using lines 68H, 168, and 410 (a metastatic subpop-

Table 1
MOLAR DRUG DOSE NEEDED TO
INHIBIT CELL GROWTH BY 50%
($\times 10^{-7}$)

	Human colon tumor subpopulation		
Drug	DLD-1	Clone A	Clone D
Methyl-CCNU	270	300	570
Vincristine	4.3	5.0	4.9
Ara C	140	730	900
5-FU	45	15	36
Actinomycin D	0.02	0.07	0.07
Mitomycin C	0.61	0.01	0.38

Data taken from References 1 to 3.

Table 2
MOLAR DRUG DOSE NEEDED TO
INHIBIT CELL GROWTH BY 50%

	Mouse mammary tumor subpopulation[a]		
Drug	68H	168	410(LM)[b]
Methotrexate ($\times 10^{-10}$)	7.0	1.4	2.9
5-FU ($\times 10^{-7}$)	3.2	0.15	4.3

[a] Data taken from Reference 4.
[b] Lung metastasis.

ulation isolated from lung). As can be seen from Table 2, there is a marked in vitro sensitivity of line 168 cells to 5-FU, while the other two tumor subpopulations are of approximately equal resistance. There is also an indication that line 68H is slightly more resistant to methotrexate than are the other two tumor lines. While data are limited, there was no apparent consistency in the in vitro and in vivo responses of these tumor lines to methotrexate or 5-FU. In vivo, at doses as high as 50 mg/kg, only line 68H showed an indication of sensitivity to methotrexate, as production of tumors in mice was reduced by approximately 50%. Neither line 168 nor 410 showed an indication of a methotrexate effect. This is clearly opposite to the effects noted in vitro. A similar response was noted for treatment with 5-FU in vivo, with line 68H showing significant inhibition of tumor production with an increased latency, whereas neither line 168 nor 410 showed any effect. This is somewhat surprising, as the in vitro data (Table 2) would suggest that line 168 should have shown marked sensitivity to the effects of 5-FU. However, these in vivo data were obtained from mice that had received subcutaneous tumor cell injections only 2 to 3 days prior to drug treatment. Therefore, the authors were not treating a neoplasm with an established vasculature.

Heppner et al.[4] also studied the effects of methotrexate and 5-FU on established tumors. Although these were very small (about 1 mm²), they should be more representative of in vivo drug effects on solid tumors. In this situation, only line 168 showed any sensitivity to methotrexate (line 68H showed no sensitivity). Also, both lines 68H and 168 showed sensitivity to 5-FU, whereas line 410 did not. These data correspond much better to the in vitro data shown in Table 2, and indicate that it is possible to obtain in vitro and in vivo data on heterogeneous subpopulations that are comparable, but that the in vivo assay must be performed when the neoplasm has been established.

An important finding by Heppner et al.[4] was that treatment of line 68H or line 168 tumors with methotrexate appeared to increase the number of metastases to the lung. This increased metastatic capability does not appear to be related directly to the intrinsic sensitivity of the subpopulations to the drug (Table 2).

In other studies, Palyi et al.[5] used the NK/Ly mouse ascites lymphoma system to study the drug sensitivity of the parent tumor line (designated P_0) vs. three clonal subpopulations (designated P_1, P_3, and P_4). In these studies, it was found that the parent line, and two subpopulations (P_4 and P_3) were very sensitive to adriamycin, while the P_1 line was about twice as resistant. In contrast, when vincristine was studied, the parent line was the most resistant line with the P_1 subpopulation being only slightly more sensitive. The P_3 and P_4 subpopulations were very sensitive to vincristine. The vincristine responses appeared to have a direct relationship to ploidy, with the more sensitive lines being hypotetraploid.

Stephens and Peacock[6] have performed an interesting study in which variants from the B16 mouse melanoma which differed in the level of expression of pigmentation were examined with respect to sensitivity to m-AMSA (4'-(9-acridinylamino)methanesulphon-m-anisidine). As m-AMSA appears to bind to melanosomes, this drug might be more effective against pigmented variants of melanoma cells. Survival curves were generated (similar in shape to X-ray survival curves) from which dose-response parameters were obtained. As such survival curves often show an inflection ("shoulder") region at low doses followed by an exponential decrease at high doses, three parameters may be defined: D_o (in units of dose), which is a measure of the cellular inactivation rate in the high dose region; n, which is a back extrapolation of the linear region of the response curve to intercept the ordinate at zero dose; and D_Q (in units of dose), which is the dose at which the back extrapolation of the linear portion of the survival curve intercepts the 100% survival level. Melanotic or amelanotic tumor clones were used to produce solid tumors which were treated with graded doses of m-AMSA, and then assayed for survival in vitro using colony formation as the endpoint. The amelanotic tumor cells gave drug survival curve parameters of n = 1.0, D_q (mg/kg) = 0.0, and D_o (mg/kg) = 12. In contrast, the highly pigmented solid tumor cells gave values of n = 9.5, D_q = 17.0, and D_o = 12. Therefore, in contrast to the original hypothesis, the melanotic clone was much more resistant to the drug (as denoted by the D_q value) than was the amelanotic clone. These results illustrate the heterogeneity in response that is obtained when tumor subpopulations are compared for any given single agent.

An interesting study has been done by Tofilon et al.[7] who used the technique of sister chromatid exchange (SCE) to detect cellular heterogeneity in the sensitivity to 1,3-Bis (2-chloroethyl)-1-nitrosourea (BCNU) in rat brain tumor cells. This assay is more sensitive than the usual clonogenic endpoint, and the authors felt that this approach was potentially a valuable addition for the evaluation of heterogeneity in drug sensitivity.

Reeve et al.[8] have investigated the patterns of cross sensitivity to nitrosoureas and nitrogen mustards in 5 clonal subpopulations from the RIF-1 mouse sarcoma. They found no correlation in responses to the two chemotherapeutic agents among the lines.

Differential responses of tumor cell subpopulations have also been demonstrated by a number of other investigators for cells from primary tumors.[9-14] Also, differences in drug responsivity have been clearly demonstrated between primary tumors and metastases,[15-19] and between different metastases in the same host.[12,20]

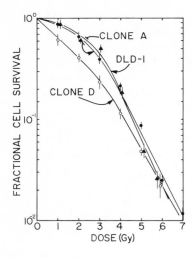

FIGURE 1. Survival response of human colon adenocarcinoma sublines (▲ DLD-1; ● clone A; ○ clone D;) to 100 kVP X-irradiation. The data are the mean survival values from replicate experiments, and the error bars are the standard errors. (From Leith, J. T., Dexter, D. L., DeWyngaert, J. K., Zeman, E. Z., Chu, M. Y., Calabresi, P., and Glicksman, A. S., Differential responses to X-irradiation of subpopulations of two heterogeneous human carcinomas in vitro, Cancer Res., 42, 2556, 1982. With permission.)

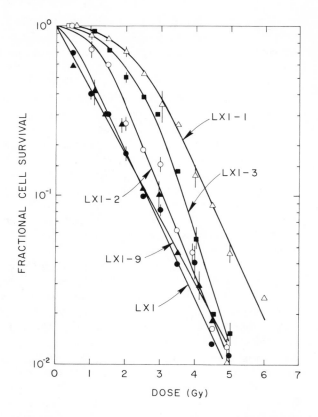

FIGURE 2. Survival response of human lung carcinoma sublines. (● LX1; △ LX1-1; ○ LX1-2; ■ LX1-3; ▲ LX1-9) to 100 kVp X-irradiation. The data are the mean survival values from replicate experiments, and the error bars are the standard errors. (From Leith, J. T., Dexter, D. L., De Wyngaert, J. K., Zeman, E. Z., Chu, M. Y., Calabresi, P., and Glicksman, A. S., Differential responses to X-irradiation of subpopulations of two heterogeneous human carcinomas in vitro, Cancer Res., 42, 2556, 1982. With permission.)

III. RESPONSES TO IONIZING RADIATION

A series of experiments dealing with the responses of two heterogeneous human carcinomas to x-irradiation in vitro have been performed by Leith et al.[21] The tumors studied were the heterogeneous DLD-1 human colon adenocarcinoma, which was cloned to provide two distinct subpopulations termed clones A and D,[1,22-24] and the heterogeneous LX1 human lung carcinoma, which was cloned to provide four distinct subpopulations (LX1-1, LX1-2, LX1-3, and LX1-9).[25-27] These tumor cell lines were grown as asynchronous, exponentially dividing cells. After irradiation, cells were replated and the survival of each tumor cell line was determined by counting the number of visible colonies that arose from single surviving cells. These data were analyzed as a function of radiation dose to obtain parameters of cellular survival as discussed previously in the section dealing with drug sensitivity (i.e., n, D_o, and D_q). The actual survival data for these tumor lines are shown in Figures 1 and 2, and the concomitant survival parameters are listed in Table 3.

There are several important features to the data shown in Figures 1 and 2, and listed in Table 3. The data illustrate that significant differences exist among tumor popula-

Table 3
RADIATION SURVIVAL PARAMETERS OF HUMAN COLON
ADENOCARCINOMA AND LUNG CARCINOMA SUBLINES

Tumor	Tumor subline	Extrapolation number	D_o(Gy)	D_q(Gy)
Colon adenocarcinoma	DLD-1(parent)	11.7	2.3	1.0
	Clone A	8.2	2.2	1.1
	Clone D	5.8	1.9	1.1
Lung carcinoma	LX1(parent)	1.2	0.2	1.1
	LX1-1	8.5	2.1	1.0
	LX1-2	2.5	0.9	1.0
	LX1-3	20.0	2.0	0.7
	LX1-9	1.0	0.0	1.1

Data taken from Leith, J. T., Dexter, D. L., DeWyngaert, J. K., Zeman, E. Z., Chu, M. Y., Calabresi, P., and Glicksman, A. S., *Cancer Res.*, 42, 2556, 1982.

tions in terms of their relative sensitivities to ionizing radiation. For the human colon adenocarcinoma, clone D is more sensitive as compared to clone A; for the human lung carcinoma, LX1 and LX1-9 are very radiosensitive, while in contrast, LX1-1 is very radioresistant. As listed in Table 3, these differences are mainly expressed in the parameter termed D_q. This value is termed the "quasithreshold dose", and its biological significance is that it describes the ability of the mammalian cell to accumulate injury before overt cell death is seen,[28,29] and is thus an index of the ability of the cell to *repair* injury after damage by ionizing radiation exposure. The target is presumed to be the DNA, and the ability to enzymatically repair damage to DNA is an important cellular survival feature. As indicated in Table 3, D_q values vary widely, from values of essentially zero (i.e., no ability to repair damage, as in lines LX1 and LX1-9) to large values (e.g., clone A, LX-1 and LX1-3). Therefore, one should expect subpopulations within a primary heterogeneous solid tumor in a patient to exhibit variability in terms of response to ionizing radiation. The parameter termed D_o (Table 3) as described previously is the rate of cellular inactivation (i.e., killing), and is defined by the linear portion of the survival curves at the higher radiation doses. The biological significance of this parameter is that it is a measure of the intrinsic sensitivity of the cell (disregarding repair capacity). While there are some differences among tumor cell responses in terms of the D_o values (Figure 1 and 2), there is much less variability in this parameter than in the D_q parameter. These results indicate that the intrinsic sensitivities (i.e., induction of damage to the target with production of a given number of radiation lesions per unit dose) of tumor subpopulations are relatively similar, but there is much diversity in the abilities of these subpopulations to *repair* such damage. Therefore, while it is possible to irradiate a heterogeneous neoplasm and create the same initial level of damage within its subpopulations, survival of these subpopulations will depend critically on the ability of each subpopulation to repair such injury.

This concept of differential repair capability after exposure to agents such as drugs or radiation among tumor subpopulations is important. First, it provides a rationale upon which therapy can be based, i.e., modification of such ability within the cancer cell to render it more sensitive. Second, in the majority of clinical applications of ionizing radiation, the doses given are typically about 2 Gy (Figure 1), and as a consequence, this will emphasize the differences in repair capacity among tumor subpopulations. At a dose level of 2 Gy it may be seen from Figure 1 that there is a factor of about 1.7 between the sensitive clone D colon line and the resistant clone A colon line in terms of relative survival. For the lung carcinoma system, this factor is even more pronounced; a fourfold difference between the most sensitive LX1-9 line and the most resistant LX1-1 line (Figure 2). Therefore, major differences in the survival of tumor subpopulations will be found at clinically relevant doses of ionizing radiation. A corollary to this, given the fact that radiation exposures are typically given in many installations (e.g., 20 to 30) is that this should impose selection pressure for radioresistant subpopulations.

Also, in both of these experimental systems, the position of the survival curve of the parent tumor line (DLD-1, LX1) cannot be predicated based on the responses of the subclones. With the colon tumor, the parent is more resistant than expected, and for the lung tumor, the parent is more sensitive than expected. While such a response may simply indicate that the full spectrum of radiation sensitivities of clonal subpopulations has not been defined because not all of the clones have been isolated, this may also indicate that clonal interactions (Chapter 7) may be an important consideration in the assessment of the overall cytotoxic responses of heterogeneous neoplasms. Also, it is important to note that subclones exist that are more radioresistant than the parent line. This must indicate that, in the progression and evolution of tumors, resistant variants to cytotoxic agents will be encountered.[30-32]

DeWyngaert et al.[33] have examined the radiation responses of four closely related sublines isolated from the mouse mammary adenocarcinoma developed by Heppner and colleagues.[4,34] The characteristics of this rodent tumor model system are further discussed in Chapters 5 and 7. In this study, the subpopulations were exposed to radiations that possess different patterns of energy deposition within the cell (X-rays and neutrons). The objectives of this work were to define the basic radiation heterogeneity of these cell lines and to investigate how such heterogeneous tumor populations respond to different types of ionizing radiation. The latter objective has implications in terms of the clinical use of ionizing radiations.

The responses of these tumor lines are shown in Figures 3 and 4 for X-ray and neutron irradiations, respectively. It is obvious that there exist significant differences in the relative effects of the two types of radiation on the different sublines. In Figure 3, line 66 is the most radiation-resistant tumor clone while line 68H is the most sensitive to x-irradiation. It is interesting that the response of line 410, which is a line derived from a metastatic deposit in the lung, is intermediate in terms of its radiation sensitivity. At the clinically relevant dose of 2 Gy, subline 66 is about 1.5 times more radiation resistant than line 68H. The data in Figure 4 for the neutron responses show that there is much less variability in responses as compared to the X-ray curves. This is due to the fact that the energy deposition produced within the DNA is much denser with neutrons than with X-rays so that the ability of the cell to repair such injury is much less. As a consequence, the survival curves tend to be very similar. This similarity suggests that irradiation with neutrons would overcome the problem of intrinsic cellular heterogeneity seen with X-rays.

Further work with both the mammary and human colon tumor model systems in terms of combined modality treatments with X-rays together with other agents is described in Section V of this chapter.

To further illustrate the heterogeneity in radiation response, it is relevant to present the studies performed by Hill and her colleagues on X-ray sensitivity of subpopulations from the B16 tumor.[35] Hill et al. obtained survival information on 10 subclones, and their data are presented in Figure 5. In this Figure, we have plotted their survival data for each of the 10 clones at doses of 2, 4, and 6 Gy, along with the mean survival of the 10 clones, and the 95% confidence limits on the response. There was a great deal of variability, again indicating that heterogeneity in response to irradiation should be expected in a heterogeneous tumor. The mean values of the survival parameters for

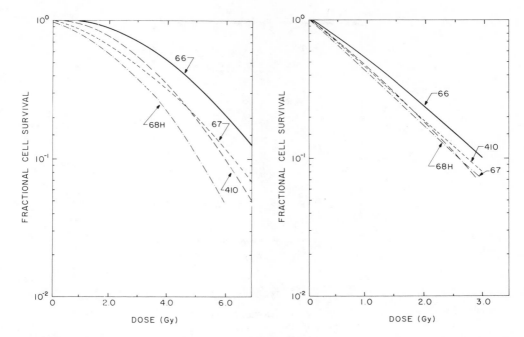

FIGURE 3. Schematic illustration of the X-ray survival curves for the mouse mammary adenocarcinoma subpopulations. (From De Wyngaert, J. K., Leith, J. T., Peck, R. A., Jr., Bliven, S. F., Zeman, E. M., Marino, S. A., and Glicksman, A. S., Differential RBE values obtained for mammary adenocarcinoma tumor cell subpopulations after 14.8 MeV neutron irradiation, *Radiat. Res.*, 88, 118, 1981. With permission.)

FIGURE 4. Schematic illustration of the neutron survival curves for the mouse mammary adenocarcinoma subpopulations.(From De Wyngaert, J. K., Leith, J. T., Peck, R. A., Jr., Bliven, S. F., Zeman, E. M., Marino, S. A., and Glicksman, A. S., Differential RBE values obtained for mammary adenocarcinoma tumor cell subpopulations after 14.8 MeV neutron irradiation, *Radiat. Res.*, 88, 118, 1981. With permission.)

these 10 clones were n = 3.1, D_o = 1.95 Gy, and D_q = 2.0 Gy. At a dose level of 2 Gy, the difference in survival between the most resistant and the most sensitive melanoma tumor population was about a factor of 2, again indicating that in a typical clinical fractionated irradiation protocol, the most resistant subpopulations would be selected.

Welch et al.[36,37] have performed an important series of experiments using the heterogeneous rat 13672NF mammary adenocarcinoma system. In these studies, the responses of both primary and metastatic clones to ionizing radiation were determined. This work provides a unique opportunity to correlate intrinsic radiation sensitivity with metastatic capacity. In Figure 6, we show data on the radiation responses of the primary tumor subpopulations at doses of up to 4 Gy. Only the MTC subpopulation appears to possess a small shoulder on its survival curve (of about 0.2 Gy), and the major difference among the subpopulations appears to be in the D_o values which are approximately 2.4., 2.1, 1.7, and 1.5 Gy for the MTC, MTF7, MTA, and MTF4 subpopulations, respectively. At a dose of 2 Gy, there is a difference in survival of about 1.6 between the most resistant and sensitive subpopulations. There is no apparent correlation between intrinsic radiation sensitivity and metastatic ability.

In Figure 7, data taken from Welch et al.[36] on the radiation responses of two metastatic subpopulations from a rat mammary adenocarcinoma are shown. The D_o values for these two tumor lines are approximately 1.8 and 1.4 Gy for the MTLn2 and MTLn3 subpopulations, respectively. There is, therefore, no apparent difference in radiosensitivity between primary and metastatic tumor lines.

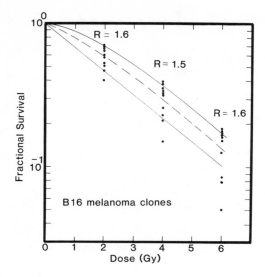

FIGURE 5. Illustration of the average survival of ten clones obtained from the B16 melanoma and treated with X-rays. The envelope indicates the 95% confidence limits on the survival responses.

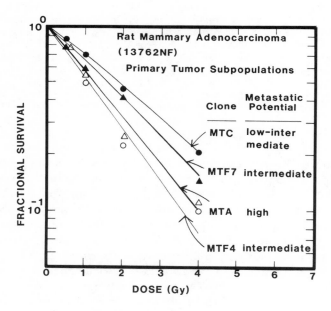

FIGURE 6. Survival responses of primary rat mammary adenocarcinoma subpopulations after γ-irradiation. Also indicated in the figure is the relative metastatic potential of each primary tumor subpopulation.

IV. RESPONSES TO HYPERTHERMIA

Elevated temperature (hyperthermia) has been used as an agent in the treatment of cancer since the early 1900s. This entails the use of either whole-body or localized hyperthermic treatments at temperatures in the range of 40 to 45°C. The potential for

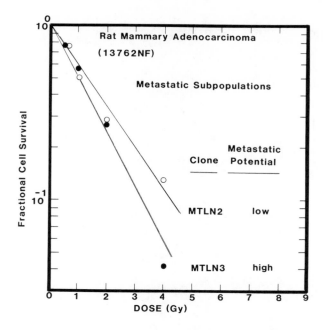

FIGURE 7. Survival responses of rat mammary adenocarcinoma metastatic subpopulations after γ-irradiation. Also indicated in the figure is the relative metastatic potential of each metastatic tumor subpopulation.

the use of this agent as either a primary or adjuvant modality in the treatment of cancer has undergone a renaissance in the last decade, due to the fact that there are several attractive features to the use of heat and due to the development of more sophisticated instrumentation. For example, the use of temperatures of about 41° or greater is cytotoxic to cells in a time-dependent fashion, and this cell killing effect can be potentiated by decreased pH as might be present within a solid tumor (Chapter 6). Also, hyperthermia is most effective in killing cells during the S phase of the cell cycle, and could be used in conjunction with agents (such as ionizing radiation) that are effective in other cell cycle phases (G_1). Hyperthermia will also kill hypoxic cells as well as oxic cells, and therefore, represents a treatment that would affect cells that might normally be resistant to certain treatments (i.e., ionizing radiation). Finally, hyperthermia has been shown to interact with certain chemotherapeutic agents to produce interactive effects. Therefore, the use of hyperthermia allows us to test another response phenotype of heterogeneous tumor subpopulations, and the sensitivities to heat can then be compared to responses for other agents to show if correlations exist.

Leith et al.[38,39] have also performed experiments on the tumor subpopulations of the DLD-1 human colon adenocarcinoma system involving hyperthermia. The thermal survival responses of the three lines comprising the DLD-1 system are shown in Figures 8 to 10. Five different temperatures were studied for each of the three tumor cell lines. In Figure 8 (42.2 to 42.5°C), it can be seen that all the survival responses are biphasic. At about 3 to 4 hr of continuous heating, the phenomenon of thermotolerance occurs, which is indicated by a change in the slope of the survival curve at longer times of heating. This phenomenon is not completely understood, but represents a real change in the responsivity of the cell to heat killing, and has been demonstrated in the majority of mammalian tumor lines. If one compares the initial (i.e., nonthermotolerant) regions of survival (0 to 3 hr), the clone D tumor line is the most resistant line to hyper-

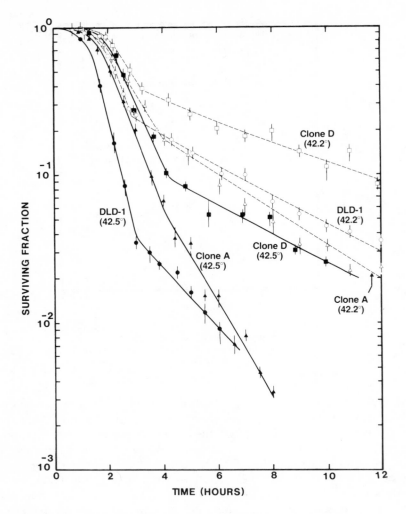

FIGURE 8. Hyperthermic survival responses for human colon adenocarcinoma lines, DLD-1, clone A, and clone D, after exposure to graded times at either 42.2° or 42.5°. The error bars are the standard errors of the mean values. (From Leith, J. T., Heyman, P., De Wyngaert, J. K., Dexter, D. L., Calabresi, P., and Glicksman, A. S., Thermal survival characteristics of cell subpopulations isolated from a heterogeneous colon tumor, *Cancer Res.*, 43, 3240, 1983. With permission.)

thermia, with clone A and DLD-1 having similar sensitivities (42.2°C). At 42.5°, responses are similar to the 42.2° responses, and the clone D line is still the most resistant line, with clone A being of intermediate sensitivity, and the parent DLD-1 line being of greatest sensitivity. It is important to consider separately the thermotolerant regions of the survival curves (3 to 12 hr). At 42.2°, clone D still exhibits the greatest hyperthermic resistance, with clone A and DLD-1 of similar sensitivities. However, at 42.5°, while clone D is again the most resistant, the DLD-1 line is of intermediate sensitivity, and clone A is the most sensitive line. It is therefore apparent that not only does the sensitivity vary among subpopulations, but that each subpopulation itself can express variability to hyperthermic treatment as a function of temperature.

Specifically, clone A, although being of intermediate sensitivity at shorter heating times, exhibits the smallest degree of induced thermotolerance. This heterogeneity can

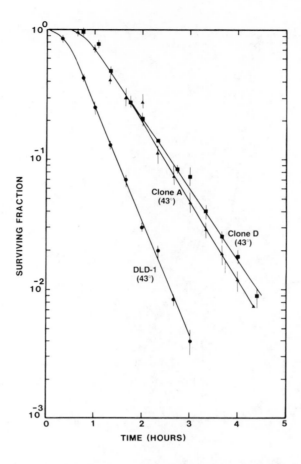

FIGURE 9. Hyperthermic survival responses for human colon adenocarcinoma lines, DLD-1, clone A, and clone D after exposure to graded times at 43°. The error bars are the standard errors of the mean values. (From Leith, J. T., Heyman, P., De Wyngaert, J. K., Dexter, D. L., Calabresi, P., and Glicksman, A. S., Thermal survival characteristics of cell subpopulations isolated from a heterogeneous colon tumor, *Cancer Res.*, 43, 3240, 1983. With permission.)

be most easily described by the thermotolerance ratio (TTR), which is the ratio of the final slope of the survival curve to the initial slope. For the DLD-1 line, this TTR is 4.9 and 3.5 for temperatures of 42.2 and 42.5, respectively, 4.3 for clone D at both temperatures, but only 2.9 and 1.5 for clone A. If induction of thermotolerance within cells during heating presents a limitation on the use of hyperthermia as a cancer modality, then it is obvious that heterogeneous tumor subpopulations will vary in this parameter. This would suggest that clone D would be an example of a very difficult tumor subpopulation to treat with hyperthermia as it is resistant at the shorter heating times, and also exhibits the greatest degree of induced thermotolerance.

In Figure 9 and 10, the survival data for these three tumor subpopulations after exposure to temperatures of 43° or 44 to 45° are shown. There is no induction of thermotolerance at these higher temperatures, due to the fact that heat killing is very rapid at 43 to 45°, so that sufficient time for the observation of induced thermotolerance is not available.

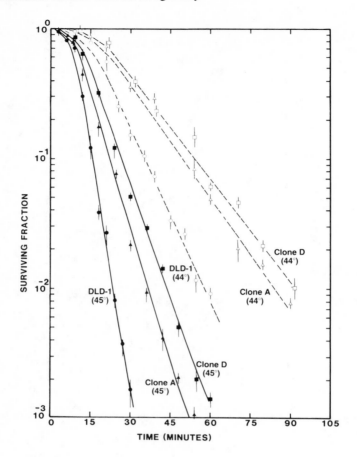

FIGURE 10. Hyperthermic survival responses for human colon adenocarcinoma lines DLD-1, clone A, and clone D after exposure to graded times at either 44° or 45°. The error bars are the standard errors of the mean values. (From Leith, J. T., Heyman, P., De Wyngaert, J. K., Dexter, D. L., Calabresi, P., and Glicksman, A. S., Thermal survival characteristics of cell subpopulations isolated from a heterogeneous colon tumor, Cancer Res., 43, 3240, 1983. With permission.)

At these temperatures, the same relative order of heat resistance is present, with clone D being the most resistant, clone A of intermediate sensitivity, and DLD-1 being the most sensitive.

These data provide insights into the responses of a heterogeneous tumor to hyperthermic killing. First, subpopulations will vary in their responses to heat. Clone D is on the average about 1.6 times more resistant than the DLD-1 line, and clone A is about 1.5 times more resistant. Second, subpopulations will vary in their expression of thermotolerance, and this variation is not apparently correlated with the heat sensitivity in the initial regions of the survival response. This suggests that the mechanisms for heat killing and heat tolerance are different, and vary in potency among neoplastic subpopulations.

Tomasovic et al.[40] and Welch et al.[37] have investigated primary and metastatic clones obtained from a rat mammary adenocarcinoma in terms of their responses to heat. These studies are similar to the work of Leith et al.[38,39] in that 42 and 45°C temperatures were investigated. The data obtained are summarized in Table 4, in which the

Table 4
SENSITIVITIES OF RAT MAMMARY TUMOR
SUBPOPULATIONS TO HYPERTHERMIA

42°C	D_o, Initial region of survival curve (hr)	D_o, Final region of survival curve (hr)	Ratio ($D_o f/D_o i$)
Primary tumor lines			
MTC	2.2	3.9	1.8
MTF$_7$	3.5	6.6	1.9
Metastatic tumor lines			
MTL$_{N2}$	3.5	7.4	2.1
MTL$_{N3}$	2.6	7.8	3.0

45°C	Single dose $D_{o\,o}$ (min)	Split dose $D_{o\,8}{}^a$ (min)	Ratio $D_{o\,8}/D_{o\,o}$
Primary tumor lines			
MTC	3.8	24.3	6.4
MTF$_7$	7.5	34.5	4.6
Metastatic tumor lines			
MTL$_{N2}$	4.2	13.8	3.3
MTL$_{N3}$	7.6	42.6	5.6

^a The subscripts on the D_o values indicate values obtained either from a single dose exposure at 45°C ($D_{o\,o}$), or from experiments in which an 8-hr interval at 37°C was given between two heat exposures ($D_{o\,8}$).

Data taken from Tomasovic, S. P., Rosenblatt, P. L., and Heitzman, D., *Int. J. Radiat. Oncol. Biol. Phys.*, 9, 1675, 1983. With permission.

inactivation rates for the different subpopulations are given. At 42°C, the survival curves were biphasic indicating the onset of thermotolerance after about 3 to 4 hr of heating (cf., Figure 7 for the shape of such curves). It is obvious from Table 4 that the two primary neoplastic subpopulations differ in both the initial and final inactivation rates for the two regions of the survival curve, although the thermotolerance ratio (final to initial slope) is approximately the same for the two tumor lines. In contrast, the two metastatic lines differ in their initial inactivation rates while having similar final slope inactivation rates. As a consequence, the thermotolerance ratios of the metastases are quite different than that of the primary neoplasms. These data again illustrate the heterogeneity that exists in terms of heat inactivation of subpopulations from either primary or metastatic tumor cells. The data also indicate that the metastatic lines cannot necessarily be expected to have heat responses similar to the primary tumor subpopulations. Tomasovic et al.[41] have also investigated the relationship between sensitivity to hyperthermic cell killing and synthesis of specific protein, termed heat shock proteins (HSP) which may be related to survival after heating, using cell clones from the rat mammary adenocarcinoma. They found a direct relationship between heat resistance and the rates of synthesis of the HSPs indicating the heterogeneity that exists in regulatory or metabolic adjustments among clonal subpopulations to this cytotoxic stress.

The data for experiments performed at 45°C further illustrate the differential responses to heat. Also, Tomasovic et al.[40] performed experiments which included an exposure to multiple heat treatments. An initial heat exposure sufficient to reduce survival to about 10% was given, the cells were returned to 37°C, and at later times, the tumor cells were reheated and the shape of the resultant survival curve determined. The interesting feature of such a fractionated exposure is that the cells develop thermotolerance during the 37°C time interval between 45°C treatments. In Table 4, their data for an 8-hr fractionated exposure as compared to cells given only a single 45°C treatment is shown. Again, heterogeneity in thermal sensitivity can be seen, for both the single and fractionated treatments, and for the primary and metastatic tumor lines. Therefore, in agreement with the data of Leith et al.[38,39] heterogeneity exists in terms of the thermal responses of both rodent and human tumor subpopulations within the range of 42 to 45°C, which is the clinically relevant region for the application of hyperthermia.

Raaphorst and Azzam[42] have published data in which the heat and radiation sensitivities of clonal subpopulations from Chinese hamster V79 cells were compared. The objectives of this work were to investigate the relationships between heat and radiation sensitivities, and to determine if hyperthermic treatment produced survivors with altered responses to subsequent heat exposure. The authors found a 12-hr treatment at 42° did not change the responses of clonal subpopulations to subsequent challenge with either heat or X-rays, indicating that heat per se apparently does not either select or induce variants with subsequently altered phenotypic properties.

In contrast, Tao et al.[43] have been able to obtain heat-resistant clones from the B-16 melanoma. Their procedure involved exposure of B-16 tumor cells to *multiple* cycles of 43° for 2.5 to 3.5 hr. Surviving clones were reexamined for their heat sensitivity, and by the 4th heating cycle, the survival of tumor cells had increased by a factor of about 100. This selection for a heat-insensitive phenotype appeared to be stable, as over 80 generations of cells from the heat resistant clones expressed the same level of heat resistance.

The authors note that this heat resistant phenotype could be demonstrated both in vitro and in vivo. As the B-16 model system allows a number of phenotypic characteristics to be intercompared (i.e., metastatic ability, lectin agglutinability, heat resistance, etc.), this system would seem to offer a powerful approach to the investigation of cellular and environmental aspects of tumor heterogeneity.

V. RESPONSES TO COMBINED MODALITY TREATMENTS

As is discussed in the preceding sections of this chapter, there is marked diversity in the responsivity of subpopulations of cells from a single primary tumor to any given cytotoxic treatment. Given this essentially "random assortment" of phenotypic responses,[30] a simple approach to the treatment of heterogeneous cancers would be the use of multiple cytotoxic agents in combination. This approach can be likened to the use of "multiple bullets" to kill a number of different targets.

For example, Leith et al.[44] have investigated the combination of hyperthermia and X-irradiation on clonal subpopulations from the heterogeneous human DLD-1 colon tumor. In this work, complete survival data were obtained for the A and D subpopulations for X-irradiation, either given alone, or combined with a 2-hr exposure to 42.5°C. The heat treatments were given either immediately before or immediately after the X-irradiations. The rationale for the investigation of the sequencing effects of hyperthermia was the possibility that heat given prior to irradiation might modify the target (DNA) so that increased numbers of lesions would be produced when the cells were irradiated, whereas hyperthermia given after irradiation might interfere with cellular repair and recovery mechanisms operating on X-ray-induced lesions. The use of the two subpopulations from the heterogeneous tumor would then allow identification of such interactive processes.

The results of these experiments are shown in Figures 11 and 12, for the clone A and clone D subpopulations, respectively. In the figures, the dashed lines indicate the limits that would be expected if each modality simply interacted in an additive manner. Supraadditivity (or true synergism) would be indicated by a response that fell outside of the limits of additivity. It may be seen that for clone A, the two modalities do interact

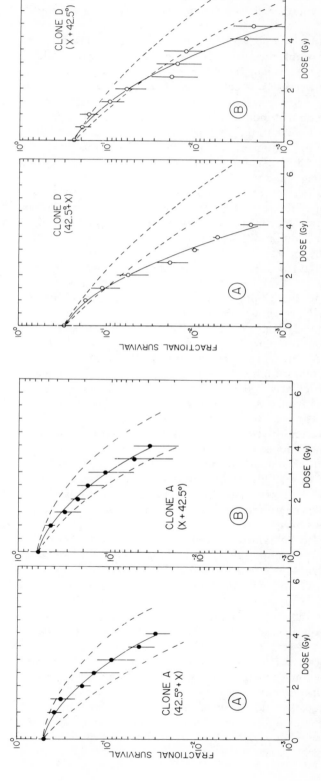

FIGURE 11. Survival responses of clone A human colon tumor cells after exposure to combined hyperthermia and X-irradiation. In (A), the responses to the combination of heat (120 min; 42.5°C) and the subsequent survival responses are shown. The survival responses include the decrement in survival produced by hyperthermia only, and the error bars are the standard errors of the main combined survivals. The dashed line represents the bondaries of additivity derived from the responses of the tumor cells to the individual modalities. (From Leith, J. T., Heyman, P., De Wyngaert, J. K., Dexter, D. L., Calabresi, P., and Glicksman, A. S., Survival responses of cell subpopulations isolated from a heterogeneous human colon tumor after combinations of hyperthermia and X-irradiation, *Int. J. Radiat. Biol.*, 43, 303, 1983. With permission.)

FIGURE 12. Survival responses of clone D human colon tumor cells after exposure to combined hyperthermia and X-irradiation. In (A), the responses to the combination of heat (120 min; 42.5° C), 3-min interval, and then X-irradiation (graded single doses) are shown. In (B), the reverse sequence of guarded single doses of X-rays followed after a 3-min interval by heat (120 min; 42.5° C) and the subsequent survival responses are shown. The survival responses include the decrement in survival produced by hyperthermia only, and the error bars are the standard errors of the main combined survivals. The dashed line represents the boundaries of additivity derived from the responses of the tumor cells to the individual modalities.

in an additive manner, and no supraadditivity exists. In contrast, for clone D, there is an indication at the higher radiation doses that true supraadditivity does exist, particularly for the sequence in which hyperthermia is given immediately prior to X-irradiation.

Therefore, these data indicate that interaction between cytotoxic modalities may be a very effective way to achieve favorable results in heterogeneous tumors. It should be appreciated that an additive response is worth achieving in its own right, and that supraadditivity (synergism) is not necessarily needed or required. It is interesting, however, that there still exists heterogeneity in response between the two clonal subpopulations even in the situation of combined modality therapy, in that clone D exhibits a more marked response than does clone A to the combination of two agents.

A different approach to the use of combined modality therapies which is somewhat novel is the use of differentiation-inducing compounds such as polar solvents, in conjunction with other agents (Chapters 3 and 9).[45-49]

As polar solvents induce a modification of the malignant phenotype of cells, it was postulated that they might also alter the response to therapeutic agents.[50] In contrast to the use of "multiple bullets", this approach essentially invokes the modification of multiple targets to a more uniform, common "target". In effect, this amounts to "homogenization" of the response phenotypes.

This hypothesis was first investigated using the mouse mammary adenocarcinoma system of Heppner and colleagues.[4,34] In this work, radiation-resistant and radiation-sensitive sublines[33,51] were grown in tissue culture for several passages in medium containing the polar solvent dimethylformamide (DMF). The monolayer cultures were irradiated, and replated into fresh tissue culture medium for assessment of cell survival using colony forming ability as the endpoint. The results obtained are listed in Figure 13, and the radiation survival parameters are listed in Table 5. The results were quite interesting, as they indicated that maturation induction was accompanied by increased radiation sensitivity. This was particularly evident in the most resistant tumor line, line 66, where the D_q value was reduced from 3.0 to 0.9 Gy, a reduction of about 60%. The more radiation-sensitive line 67 was not affected to the same extent, but as line 67 had a small value of D_q to start with, a major reduction in this value might not have been expected. This finding suggests that induced differentiation is accompanied by a reduction in the ability to repair radiation damage produced in the cell. If maturational agents were to selectively act against cells possessing the largest repair capacities (i.e., the largest D_q values), this would be of significant potential as a therapy in the treatment of heterogeneous tumors, as it might reduce the heterogeneity of subpopulations to a homogeneous, more radiation-sensitive state.

To extend these initial findings to human tumor cells, Leith et al.[52] studied the responses of the clone A and clone D subpopulations of the human DLD-1 colon adenocarcinoma after x-irradiation combined with dimethylformamide treatment. The same experimental protocol was followed with the human carcinoma cells as with the mouse mammary carcinoma cells. The results are shown in Figure 14 and 15 and are listed in Table 6. Additionally, irradiated cells were also placed back into medium containing DMF as one of the experimental conditions. The objective of this was to determine if DMF also would interfere with postirradiation recovery from radiation injury as well as modify the cell prior to irradiation. As can be seen from Table 6, non-DMF treated tumor cells irradiated and placed into DMF-containing medium showed essentially no modification of their inherent radiation sensitivity parameters. However, DMF pretreatment did modify the radiation survival parameters, decreasing the D_q values in clone A and clone D cells by about 20%. This could be increased further if DMF-pretreated and -irradiated cells were placed back into medium containing DMF after irradiation. In this case, the D_q values were reduced by about 50% in clone A,

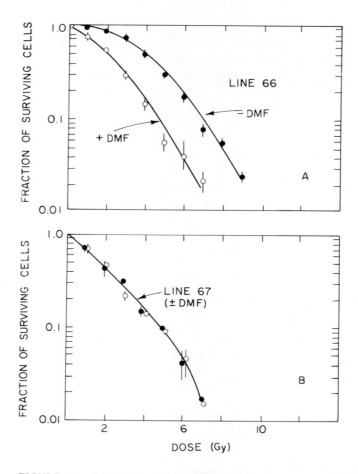

FIGURE 13. Fractional survival of line 66 (panel A) and line 67 (panel B) mammary adenocarcinoma tumor cells in vitro. Cells were irradiated (100 kVp X-rays) either in the presence or absence of 0.8% N,N-dimethylformamide (DMF). Error bars are the 95% confidence limits on the mean survival values.(From Leith, J. T., Brenner, H. J., De Wyngaert, J. K., Dexter, D. L., Calabresi, P., and Glicksman, A. S., Selective modification of the X-ray survival response of two mammary adenocarcinoma sublines by N,N-dimethylformamide, Int. J. Radiat. Oncol. Biol. Phys., 7, 943, 1981. With permission.)

and by about 40% in clone D. This finding suggests that if maturational therapy of heterogeneous tumors were to be used in conjunction with ionizing radiation, it should be administered for a time period preirradiation sufficient to induce maturation, and that DMF administration should be continued after irradiation to interfere with postirradiation recovery processes. Also, these experiments indicate that the phenomenon of radiation sensitization occurs in *both* human and rodent tumor cells after treatment with polar solvents.

It is important to note that in addition to the apparent sensitizing effects on tumor cells by combinations of ionizing radiation and polar solvents, Dexter et al.[53] have demonstrated that polar solvents will also increase the sensitivity of cultured human colon cancer cells to *cis*-platinum and mitomycin C. In this work using growth inhibition, both DMF and its metabolite N-methylformamide (NMF) were studied. It was found that DMF increased the sensitivity of colon tumor cells to *cis*-platinum by a

Table 5
MODIFICATION OF RADIATION
SURVIVAL PARAMETERS OF MOUSE
MAMMARY ADENOCARCINOMA
TUMOR SUBPOPULATIONS BY
N'N- DIMETHYLFORMAMIDE

Tumor subpopulation	Extrapolation number	$D_Q(Gy)$	$D_o(Gy)$
66 (−DMF)	6.4	3.0	1.6
(+DMF)	1.8	0.9	1.6
67 (−DMF)	2.7	1.4	1.4
(+DMF)	1.8	0.9	1.6

From Leith, J. T., Brenner, H. J., DeWyngaert, J. K., Dexter, D. L., Calabresi, P., and Glicksman, A. S., *Int. J. Radiat. Oncol. Biol. Phys.*, 7, 943, 1981.

factor of 2 to 4, and to mitomycin C by a factor of 3 to 4. Also, NMF increased sensitivity to *cis*-platinum by a factor of 5 to 10, and to mitomycin C by a factor of 3 to 4. Therefore, both polar solvents produce an increased response to chemotherapeutic agents, again indicating that maturational therapy may be an efficacious combination for the clinical treatment of heterogeneous cancers.

VI. RELATIVE RANKINGS OF RESPONSES TO VARIOUS CYTOTOXIC AGENTS

A major point in research dealing with the effects of multiple cytotoxic modalities on tumor subpopulations is the question of possible correlations. For example, the rank ordering of decreasing sensitivity of the human colon tumor lines to heat is DLD-1, clone A, and clone D (Figures 8 to 10). Also, these lines have been examined with regard to their X-ray sensitivity (Figure 1, Table 3), and there the clone D line was the most sensitive, with DLD-1 being of intermediate sensitivity and clone A being the most resistant line. There is, therefore, no correlation between heat and radiation in terms of the effectiveness of these agents on the subpopulations studied. Similar lack of correlation has been found from studies on various chemotherapeutic agents in the DLD-1 systems.[1] Indeed, the only agent which to date has exhibited a rank ordering similar to that for hyperthermic inactivation has been actinomycin D.[2] It would be of some importance if the responses of heterogeneous tumor subpopulations could be shown to exhibit correlations in terms of relative responses to cytotoxic agents, as this would directly impinge on concepts in the design of experimental and clinical therapy, i.e., the question of using multiple agents vs. the search for a common target. A discussion of such "commonalities" is presented in Chapter 9.

Additionally, the data of Tomasovic et al.[40] (Figures 5 and 6 and Table 4), also showed that there was no correlation between the sensitivities of the various rat mammary adenocarcinoma lines to X-rays or to heat.

Finally, Rofstad and Brustad[54] have examined the differential responses to X-irradiation and hyperthermia in five clonal lines from a single human melanoma in vitro. Again, there was no apparent correlation between X-ray and heat sensitivities among these five lines.

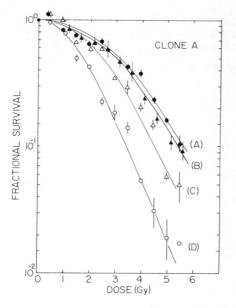

FIGURE 14. Survival responses of clone A human colon carcinoma cells after X-irradiation. Curve A, response of irradiated control cells; curve B, response of control cells irradiated and recultured in complete fresh medium containing DMF (0.8%); curve C, response of cells grown prior to irradiation in DMF (0.8%), irradiated, and recultured in complete fresh medium; curve D, response of cells grown prior to irradiation in DMF (0.8%). The error bars are the standard errors of the means. (From Leith, J. T., Gaskins, L. A., Dexter, D. L., Calabresi, P., and Glicksman, A. S., Alteration of the survival response of two human colon carcinoma subpopulations to X-irradiation by N,N-dimethylformamide, *Cancer Res.*, 42, 30, 1982. With permission.)

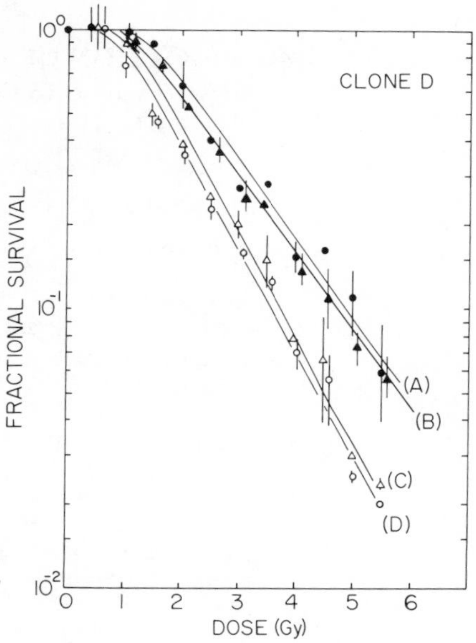

FIGURE 15. Survival responses of clone D human colon carcinoma cells after X-irradiation. Curve A, response of irradiated controlled cells; curve B, response of control cells irradiated and recultured in complete fresh medium containing DMF (0.8%); curve C, response of cells grown prior to irradiation in DMF (0.8%), irradiated, and recultured in complete fresh medium; curve D, response of cells grown prior to irradiation in DMF (0.8%), irradiated, and recultured in complete fresh medium containing DMF (0.8%). The error bars are the standard errors of the means. (From leith, J. T., Gaskins, L. A., Dexter, D. L., Calabresi, P., and Glicksman, A. S., Alteration of the survival response of two human colon carcinoma subpopulations to X-irradiation by N,N-dimethylformamide, *Cancer Res.*, 42, 30, 1982. With permission.)

VII. SUMMARY

We have attempted in this chapter to present data concerning the effects of various agents, including both cytotoxic and maturational treatments, on the responses of heterogeneous neoplasms. We have mainly addressed effects on primary tumors, but data on metastatic lesions have been included where they would illuminate the discussion. In particular, the data from investigators who have examined the effects of multiple agents within the context of a given heterogeneous model tumor system have provided much insight. Experiments involving the applications of drugs, ionizing radiation, hyperthermia, and maturational agents either alone or in combination have allowed us to reach the following set of conclusions:

A. Distribution of Responsivity within Heterogeneous Tumors

From the data reported in mouse, rat, and human tumor model systems, it is quite apparent that a distribution in the sensitivity of a given response phenotype will be

Table 6
MODIFICATION OF RADIATION SURVIVAL
PARAMETERS OF HUMAN COLON
TUMOR SUBPOPULATIONS BY N,N-
DIMETHYLFORMAMIDE

Tumor subpopulation	Extrapolation number	$D_Q(Gy)$	$D_o(Gy)$
Non-DMF treated cells			
Clone A			
Medium without DMF	6.3	2.5	1.4
Medium with DMF	4.9	2.2	1.4
Clone D			
Medium without DMF	2.3	1.3	1.6
Medium with DMF	2.2	1.1	1.5
DMF pretreated cells			
Clone A			
Medium without DMF	5.2	1.9	1.2
Medium with DMF	2.8	1.1	1.0
Clone D			
Medium without DMF	2.2	1.0	1.2
Medium with DMF	2.0	0.8	1.2

Data taken from Leith, J. T., Gaskins, L. A., Dexter, D. L., Calabresi, P., and Glicksman, A. S., *Cancer Res.,* 42, 30, 1982.

obtained for any single agent employed. Both sensitive and resistant subpopulations will be present. While sensitive subpopulations are likely not to be a clinical problem (with respect to the single agent used), the resistant subpopulations are a problem, and it is vital to consider how such resistance corresponds to the tolerance of the surrounding normal tissues for the therapeutic agent employed, as this will limit the application of that agent.

B. Selection of Resistant Subpopulations

It is also apparent that the use of single agents given in a multiple series of exposures will select the most resistant subpopulations for that given agent. Such resistant subpopulations should dominate the tumor at the end of a course of single agent therapy. In this regard, the section in Chapter 7 dealing with the potential reestablishment of equilibrium conditions in perturbed heterogeneous tumors is relevant. To date, no studies have been done examining the return to pretreatment conditions in a perturbed heterogeneous human tumor system. It should not be forgotten that most cytotoxic agents are also mutagenic, and as such may create additional resistant populations through mutational events in target cells.

C. Random Association of Response Phenotypes

The correlation of sensitivity or resistance of subpopulations to different cytotoxic modalities appears to be essentially unpredictable. This unpredictable hierarchy of phenotypic response, therefore, does not allow us to make any *a priori* predictions with regard to combinations of agents that might preferentially be given together to create an additive or supraadditive effect if the "multiple bullet" approach is taken. Conversely, we do not know which agents are contraindicated because they attack the same phenotypic target in a multiphenotype neoplasm, and are therefore ineffective in combination.

D. Clinical Approach to Heterogeneous Tumors

As stated previously, one obvious approach then is to continue to follow, in an empirical manner, the clinical approach of using combinations of agents or modalities with the hope that there is enough diversity among agents that they will effectively act upon different response phenotypes of heterogeneous tumors with a positive effect. In this approach, any supraadditivity or "synergism" is an unexpected bonus. This is exactly the basis upon which agents are currently either added or deleted to clinical protocols. It is unfortunate to note that at present, no more effective strategies based on biological properties of heterogeneous tumors are available. However, it is entirely reasonable to expect that such insight will be attained given further studies on heterogeneous systems.

However, as we have previously indicated, another potential approach would be to try to make all targets of essentially equal sensitivity to a given therapeutic modality. This is a rationale for the use of differentiating or maturational compounds in cancer therapy. Indeed, by "homogenizing" the targets, it may be possible to reach a lowest common denominator of response not only because of an increased sensitivity of the targets, but also because fewer "bullets" will be needed. This concept of "commonalities" that may be exploitable in the therapy of heterogeneous neoplasms is advanced further in Chapter 9.

Other important questions in the discussion of responses of heterogeneous neoplasms to therapy (also discussed in Chapter 5) are "how many subpopulations exist", and "what is the approximate volume percentage of each subpopulation within a heterogeneous tumor?"

The problem is further complicated by the well-known fact that solid tumors are not only heterogeneous in terms of cellular composition, but also are heterogeneous in terms of their environmental status (e.g., hypoxic regions, etc.). This microenvironmental heterogeneity will further complicate the treatment of heterogeneous solid tumors in terms of such considerations as drug distribution, alteration of radiation sensitivity, etc. (Chapter 6). Therefore, when tumor heterogeneity is considered, the concept of multiple levels of diversity which overlap should always be kept in mind.

Finally, the discussion presented in this chapter has been, by design, somewhat artifical in the sense that we have been addressing responses of the primary tumor as the important endpoint. We realize (Chapter 9) that the most important endpoint in regard to the ultimate curability of a tumor and the survival of the patient with cancer is whether one can effectively deal with the metastatic aspects of the disease, as the majority of tumors have already spread by the time of clinical presentation. Still, the four conclusions listed at the beginning of this section also hold true with respect to the consideration of (heterogeneous) metastatic disease, and strategies that are effective against primary tumors should also be effective against metastatic lesions. It may be that the concept of "commonalities" of heterogeneous tumors, which have yet to be completely defined, will ultimately yield the Rosetta stone to the problem of tumor heterogeneity (Chapter 9). Much research will have to be performed before this becomes clear.

REFERENCES

1. Calabresi, P., Dexter, D. L., and Heppner, G. H., Clinical and pharmacological implications of cancer cell differentiation and heterogeneity, *Biochem. Pharmacol.*, 28, 1933, 1979.
2. Dexter, D. L., Fligiel, Z., Vogel, R., and Calabresi, P., Heterogeneity of neoplastic cells from a single human colon carcinoma, *Proc. Am. Assoc. Cancer Res.*, 20, 199, 1979.

3. Dexter, D. L., Spremulli, E. N., Fligiel, Z., Barbosa, J. A., Vogel, R., Van Voorhees, A., and Calabresi, P., Heterogeneity of cancer cells from a single human colon carcinoma, *Am. J. Med.*, 71, 949, 1981.
4. Heppner, G. H., Dexter, D. L., DeNucci, T., Miller, F. R., and Calabresi, P., Heterogeneity in drug sensitivity among tumor cell subpopulations of a single mammary tumor, *Cancer Res.*, 38, 3758, 1978.
5. Palyi, I., Olah, E., and Sugar, J., Drug sensitivity studies on clonal cell lines isolated from heteroploid tumour cell populations. I. Dose response of clones growing in monolayer cultures, *Int. J. Cancer*, 19, 859, 1977.
6. Stephens, T. C. and Peacock, J. H., Clonal variation in the sensitivity of B16 melanoma to m-AMSA, *Br. J. Cancer*, 45, 821, 1982.
7. Tofilon, P. J., Wheeler, K. T., and Deen, D. F., Detection of heterogeneity in the chemosensitivity of 9L brain tumor cell lines to 1,3-Bis(2-chloroethyl)-1-nitrosourea by the sister chromatid exchange assay, *Eur. J. Cancer Clin. Oncol.*, 20, 927, 1984.
8. Reeve, J. G., Wright, K. A., and Twentyman, P., Patterns of cross-sensitivity in the responses of clonal subpopulations isolated from the RIF-1 mouse sarcoma to selected nitrosoureas and nitrogen mustards, *Br. J. Cancer*, 50, 153, 1984.
9. Barranco, S. C., Drewinko, B., and Humphrey, R. M., Differential responses by human melanoma cells to 1,3-bis-(2-Chloroethyl)-1-nitrosourea and bleomycin, *Mutat. Res.*, 19, 277, 1973.
10. Barranco, S. C., Ho, D., Drewinko, B., Romsdahl, M. M., and Humphrey, R. M., Differential responses by human melanoma cells grown *in vitro* to arabinosylcytosine, *Cancer Res.*, 32, 2733, 1972.
11. Hakansson, L. and Trope, C., On the presence within tumors of clones that differ in sensitivities to cytostatic drugs, *Acta Pathol. Microbiol. Scand. (A)*, 82, 35, 1974.
12. Siracky, J., An approach to the problem of heterogeneity of human tumour cell populations, *Br. J. Cancer*, 39, 570, 1979.
13. Stephens, T. C., Adams, K., and Peacock, J. H., Identification of a subpopulation of MeCCNU resistant cells in previously untreated Lewis lung tumours, *Br. J. Cancer*, 50, 77, 1984.
14. Smith, K. A., Begg, A. C., and Denekamp, J., Differences in chemosensitivity between subcutaneous and pulmonary tumours, *Eur. J. Cancer Clin. Oncol.*, 21, 249, 1985.
15. Slack, N. H. and Bross, I. D. J., The influence of site of metastasis on tumour growth and response to chemotherapy, *Br. J. Cancer*, 32, 78, 1975.
16. Trope, C., Different sensitivity to cytostatic drugs of primary tumor and metastasis of the Lewis carcinoma, *Neoplasma*, 22, 171, 1975.
17. Fugmann, R. A., Anderson, J. C., Stolfi, R. L., and Martin, D. S., Comparision of adjuvant chemotherapeutic activity against primary and metastatic spontaneous murine tumors, *Cancer Res.*, 37, 496, 1977.
18. Lotan, R. and Nicolson, G. L., Heterogeneity in growth inhibition by B-transretinoic acid of metastatic B16 melanoma clones and *in vivo*-selected cell variant lines, *Cancer Res.*, 39, 4767, 1979.
19. Biorklund, A., Hakansson, L., Stenstarn, B., Trope, C., and Ackerman, M., Heterogeneity of non-Hodgkin's lymphomas as regards sensitivity to cytostatic drugs, *Eur. J. Cancer*, 16, 647, 1980.
20. Tsuruo, T. and Fidler, I. J., Differences in drug sensitivities among tumor cells from parental tumors, selected variants and spontaneous metastases, *Cancer Res.*, 41, 3058, 1981.
21. Leith, J. T., Dexter, D. L., DeWyngaert, J. K., Zeman, E. Z., Chu, M. Y., Calabresi, P., and Glicksman, A. S., Differential responses to X-irradiation of subpopulations of two heterogeneous human carcinomas *in vitro*, *Cancer Res.*, 42, 2556, 1982.
22. Calabresi, P., Cancer: crab or chimera? The clinical implications of cancer cell heterogeneity, *Trans. Am. Clin. Climatol. Assoc.*, 92, 49, 1980.
23. Dexter, D. L., Barbosa, J. A., and Calabresi, P., N,N-Dimethylformamide-induced alteration of cell culture characteristics and loss of tumorigenicity in cultured human colon carcinoma cells, *Cancer Res.*, 39, 1020, 1979.
24. Tibbets, L. M., Chu, M. Y., Hager, J. C., Dexter, D. L., and Calabresi, P., Chemotherapy of cell-line derived human colon carcinomas in mice immunosuppressed with antithymocyte serum, *Cancer (Philadelphia)*, 40, 2651, 1977.
25. Calabresi, P. and Dexter, D. L., Clinical implications of cancer cell heterogeneity, in *Tumor Cell Heterogeneity*, 4th Bristol-Meyers Symp. on Cancer Research, Owens, A. H., Ed., Academic Press, New York, 1982, 181.
26. Chu, M. Y., Dexter, D. L., Melvin, J. B., Robison, B. S., Parks, R. E., Jr., and Calabresi, P., 2'-Deoxycoformycin in combination with 8-azaadenosine against human carcinomas and leukemia, *Proc. Am. Assoc. Cancer Res.*, 21, 269, 1980.
27. Chu, M. Y., Takeuchi, T., Yeskey, K. S., Bogaars, H., and Calabresi, P., Tumor cell heterogeneity in human lung carcinoma LX-1, *Proc. Am. Assoc. Cancer Res.*, 20, 151, 1979.

28. Elkind, M. M. and Sutton, H., X-ray damage and recovery in mammalian cells in culture, *Nature (London)*, 184, 1293, 1959.
29. Whitmore, G. F. and Gulyas, S., Studies on recovery processes in mouse L cells, *Natl. Cancer Inst. Monogr.*, 24, 14, 1967.
30. Foulds, L., The experimental study of tumor progression: a review, *Cancer Res.*, 14, 337, 1954.
31. Nowell, P. C., The clonal evolution of tumor cell populations, *Science*, 194, 23, 1976.
32. Goldie, J. H. and Coldman, A. J., A mathematic model for relating the drug sensitivity of tumors to their spontaneous mutation rate, *Cancer Treat. Rep.*, 63, 1727, 1979.
33. DeWyngaert, J. K., Leith, J. T., Peck, R. A., Jr., Bliven, S. F., Zeman, E. M., Marino, S. A., and Glicksman, A. S., Differential RBE values obtained for mammary adenocarcinoma tumor cell subpopulations after 14.8 MeV neutron irradiation, *Radiat. Res.*, 88, 118, 1981.
34. Dexter, D. L., Kowalski, H. M., Blazar, B. A., Fligiel, Z., Vogel, R. L., and Heppner, G. H., Heterogeneity of tumor cells from a single mouse mammary tumor, *Cancer Res.*, 38, 3174, 1978.
35. Hill, H. Z., Hill, G. J., Miller, C. F., Kwong, F., and Purdy, J., Radiation and melanoma: response of B16 tumor cells and clonal lines to *in vitro* irradiation, *Radiat. Res.*, 80, 259, 1979.
36. Welch, D. R., Milas, L., Tomasovic, S. P., and Nicolson, G. L., Heterogeneous response and clonal drift of sensitivities of metastic 136762NF mammary adenocarcinoma clones to radiation *in vitro*, *Cancer Res.*, 43, 6, 1983.
37. Welch, D. R., Evans, D. P., Tomasovic, S. P., Milas, L., and Nicolson, G. L., Multiple phenotypic divergence of mammary adenocarcinoma cell clones, II. Sensitivity to radiation, hyperthermia, and FUdR, *Clin. Exp. Metastasis*, 2, 357, 1984.
38. Leith, J. T., DeWyngaert, J. K., Dexter, D. L., Calabresi, P., and Glicksman, A. S., Differential sensitivity of three human colon adenocarcinoma lines to hyperthermic cell killing, *Natl. Cancer Inst. Mongr.*, 61, 381, 1982.
39. Leith, J. T., Heyman, P., DeWyngaert, J. K., Dexter, D. L., Calabresi, P., and Glicksman, A. S., Thermal survival characteristics of cell subpopulations isolated from a heterogeneous human colon tumor, *Cancer Res.*, 43, 3240, 1983.
40. Tomasovic, S. P., Rosenblatt, P. L., and Heitzman, D., Heterogeneity in induced thermal resistance of rate tumor cell clones, *Int. J. Radiat. Oncol. Biol. Phys.*, 9, 1675, 1983.
41. Tomasovic, S. P., Rosenblatt, P. L., Johnston, D. A., Tang, K., and Lee, P. S. Y., Heterogeneity in induced heat resistance and its relation to synthesis of stress proteins in rat tumor cell clones, *Cancer Res.*, 44, 5850, 1984.
42. Raaphorst, G. P. and Azzam, E. I., Heat and radiation sensitivity of Chinese hamster V79 cells and of nine clones selected from survivors of a thermal tolerant cell population, *Int. J. Radiat. Oncol. Biol. Phys.*, 6, 1577, 1980.
43. Tao, T.-W., Calderwood, S., and Hahn, G. W., Stable heat-resistant clones selected from wild-type and surface variants of B-16 melanoma, *Int. J. Cancer*, 32, 533, 1983.
44. Leith, J. T., Heyman, P., DeWyngaert, J. K., Dexter, D. L., Calabresi, P., and Glicksman, A. S., Survival responses of cell subpopulations isolated from a heterogeneous human colon tumour after combinations of hyperthermia and X-irradiation, *Int. J. Radiat. Biol.*, 43, 303, 1983.
45. Avdalovic, N. and Aden, D., Bromodeoxyuridine-(BrdUrd) and dimethylformamide-(DMF) induced changes in the surface of cultured hamster melanoma cells, *Proc. Am. Assoc. Cancer Res.*, 19, 195, 1978.
46. Borenfreund, E., Steinglass, M., Korngold, G., and Bendich, A., Effect of dimethylsulfoxide and dimethlformamide on the growth and morphology of tumor cells, *Ann. N.Y. Acad. Sci.*, 243, 164, 1975.
47. Collins, S. J., Ruscetti, F. W., Gallagher, R. E., and Gallo, R. C., Terminal differentiation of human promyelocytic leukemia cells induced by dimethylsulfoxide and other polar compounds, *Proc. Natl. Acad. Sci. U.S.A.*, 75, 2458, 1978.
48. Hager, J. C., Gold, D. V., Barbosa, J. A., Fligiel, Z., Miller, F., and Dexter, D. L., N,N-Dimethylformamide-induced modulation of organ- and tumor-associated markers in cultured human color carcinoma cells, *J. Natl. Cancer Inst.*, 64, 439, 1980.
49. Scher, W., Preisler, H. D., and Friend, C., Hemoglobin synthesis in murine virus-induced leukemic cells *in vitro*: III. Effects of 5-bromo-2'-deoxyuridine, dimethylformamide and dimethylsulfoxide, *J. Cell. Physiol.*, 81, 62, 1973.
50. Dexter, D. L., Leith, J. T., Crabtree, G. W., Parks, R. E., Jr., Glicksman, A. S., and Calabresi, P., N,N-Dimethylformamide-induced modulation of responses of tumor cells to conventional anti-cancer treatment modalities, in *Maturation Factors and Cancer*, Moore, M. A. S., Ed., Raven Press, New York, 1981, 105.
51. Leith, J. T., Brenner, H. J., DeWyngaert, J. K., Dexter, D. L., Calabresi, P., and Glicksman, A. S., Selective modification of the X-ray survival response of two mammary adenocarcinoma sublines by N,N-dimethylformamide, *Int. J. Radiat. Oncol. Biol. Phys.*, 7, 943, 1981.

52. Leith, J. T., Gaskins, L. A., Dexter, D. L., Calabresi, P., and Glicksman, A. S., Alteration of the survival response of two human colon carcinoma subpopulations to X-irradiation by N,N-dimethylformamide, *Cancer Res.*, 42, 30, 1982.
53. Dexter, D. L., DeFusco, D. J., McCarthy, K., and Calabresi, P., Polar solvents increase the sensitivity of cultured human colon cancer cells to cisplatinum and mitomycin-C, *Proc. Am. Assoc. Cancer Res.*, 24, 267, 1983.
54. Rofstad, E. K. and Brustad, T., Differential responses to radiation and hyperthermia of cloned lines from a single human melanoma xenograft, *Int. J. Radiat. Oncol. Biol. Phys.*, 10, 857, 1984.

Chapter 9

THERAPY OF METASTATIC CANCERS

I. INTRODUCTION

The ocurrence of metastasis in a cancer patient signals a critical progression in that individual's disease.[1] Successful local control no longer indicates a successful outcome; the malignancy has gone beyond the reach of the surgeon's knife. The fate of the patient now depends on how successfully this disseminated disease can be treated. Unfortunately, the statistics indicate only too clearly how often clinicians fail to cure these individuals. Approximately 70% of individuals with lung cancer have documentable disease in regional nodes or distant sites at the time their malignancy is diagnosed.[2] Small cell lung cancer is now considered to be disseminated at diagnosis; surgical intervention is not considered efficacious with these patients who are consequently treated with chemotherapy and/or radiation therapy. Clinically evident hematogenous dissemination also occurs in about one third of patients with colorectal carcinoma at presentation, and many more have micrometastases.[3] Only a few individuals (about 2%) with pancreatic carcinoma are cured by local control procedures; most of these patients die from liver metastases.[4] Approximately 50% of women with breast cancer have positive axillary nodes at diagnosis, indicating the probability of further disseminated disease.[5]

This chapter will focus on the treatment of metastatic disease; the process itself was discussed in Chapter 4. The key to developing any useful approaches to therapy is to gain an appreciation of the role of tumor heterogeneity in the occurrence and eradication of metastases. This role will be examined from both experimental and clinical perspectives.

II. EXPERIMENTAL MODELS OF METASTASIS

Research on the treatment of metastases has been dependent on the availability of appropriate laboratory models. Work with murine tumor cells has proceeded according to two principal strategies, isolation of metastases from spontaneously occurring animal neoplasms, and selection of cells with metastatic potential from mouse tumors propagated in syngeneic hosts or as cultured cell lines. These strategies will be reviewed here, and the responses of metastatic cells to therapeutic modalities will be discussed. Investigations in the past 10 years have demonstrated clearly that the progressed, disseminating murine cancer cell arises in a heterogeneous primary tumor or cell line (see Chapter 4 for a discussion). Therefore, all results obtained with such cells must be assessed with the understanding that intraneoplastic diversity provides a conceptual basis for interpretation of these data.

One of the best studied models of tumor heterogeneity is the mouse mammary carcinoma system developed by Heppner and her colleagues. Three of the original four subpopulations isolated by this group were obtained from a cell line established from a primary mouse mammary tumor using various cloning techniques. The fourth subpopulation was established from a lung metastasis from the tenth generation passage of the original tumor. This subpopulation responded in vivo to anticancer drugs differently from the primary carcinoma-derived clones, as will be discussed below.[6]

Many of the data on the responses of cells with metastatic capability to drugs or radiation were obtained in studies using selected tumor lines with aggressive potential. A variety of selection and isolation procedures have been employed; these often have been quite clever in their design. As has been described in Chapter 4, cloning a heterogeneous B16 melanoma cell line in soft agar has yielded clones with high metastatic potential.[7] Innovative techniques developed by Poste et al. (also Hart) have resulted in the selection of highly invasive cell lines.[8,9] In these experiments, animal organs or tissues such as the mouse bladder or the dog vein have been employed as a barrier, which B16 melanoma cells must actively penetrate to reach the other side. The tumor cells are deposited inside the tissue receptacle, and invading cells are collected in the soft agar milieu (or appropriate device) surrounding the vessel. Repeated passage through the bladder, for example, has selected the variant BL6 melanoma subpopulation, which is considerably more invasive than the original line.[8,9]

Other investigations which have resulted in the successful selection of metastatic variants from heterogeneous tumors or cell lines include the following. Reading et al. employed immobilized lectin to isolate cells from the murine lymphosarcoma RAW117-P cell line, which had a significantly increased capability of forming liver metastases following i.v. inoculation.[10] Nicolson et al. used brain or ovarian colonization of B16 cells in multiple cyclings of these site-specific metastases through tissue culture and mice.[11] In this way, they were able to isolate variants with an affinity for dissemination to brain or ovary upon i.v. injection into syngeneic mice. Miner et al. have also used this strategy to isolate B16 variants with increased brain colonizing ability following tail vein inoculation.[12] Talmadge and Fidler obtained cells from spontaneous metastases of four s.c. primary heterogeneous murine tumors, the UV-2237 fibrosarcoma, the K-1735 melanoma, the M5076 reticulum cell sarcoma, and the Lewis lung carcinoma. Cells from metastases from each of these tumors had enhanced metastatic potential compared to the heterogeneous parent lines that produced the primary s.c. neoplasms.[13]

Dexter et al. have utilized the chicken choriollantoic membrane (CAM) to select for metastatic variants of the B16 melanoma and the C6 rat glioma.[14] Tumor cells were deposited on CAMs of eggs 10 days postfertilization. After hatching 10 days later, the chickens were autopsied and organs were removed, minced, and implanted s.c. in C57BL mice (for melanoma) or nude mice (for glioma). Using this outgrowth strategy, a melanoma developed from a chick liver fragment, and a glioma grew out from a chick lung implant. Moreover, the s.c. melanoma metastasized to the lung of its C57BL host, and the glioma metastasized to the liver of the nude mouse recipient. Cell lines were established from the s.c. glioma and melanoma, and from a liver metastasis of the s.c. glioma. These lines were used in subsequent drug sensitivity studies.[14]

Metastatic cells from human tumors have not been as readily obtainable. Tsuruo and Fidler isolated lung metastases from nude mice that had received tail vein injection of A-375 human melanoma cells at 3 weeks of age.[15] It was postulated that the cells could clone the lungs of young nude mice because the natural killer (NK) cell level was still low in 3-week-old mice.[16] Such experiments have in general proven unsuccessful with older athymic mice, due presumably to a high level of NK cells. Dexter et al. have isolated two distinct subpopulations (clone A and clone D) from a heterogeneous primary human colon carcinoma DLD-1.[17,18] Spremulli et al. have demonstrated that clone A has greater metastatic potential than clone D in the nude mouse, and cell lines have been established from the hematogeneous clone A secondary tumors.[19] A number of studies have been performed utilizing clones A and D as targets for drugs, radiation, and hyperthermia; these have been discussed in Chapter 8.

III. LABORATORY STUDIES ON THE SENSITIVITY OF METASTATIC CELLS TO CHEMOTHERAPEUTIC AGENTS

Chapter 8 has reviewed the heterogeneity displayed by tumor cell subpopulations in their responses to treatment modalities. Most of the early work documenting differential chemosensitivities among clones from a single neoplasm studied subpopulations from primary tumors. However, a number of investigations focusing on the responses of metastases compared to the primary tumor or to other metastases from that tumor have also been reported. These studies have utilized metastatic cell lines or metastases that were isolated by selection methods described above.

Dexter et al. have demonstrated that chicken chorioallantoic membrane (CAM)-selected variant C6 glioma cells were 33-fold more sensitive to vincristine than were the parent glioma cells.[14] Heppner et al. compared the response of the 4.10 lung metastasis-derived mouse mammary tumor subpopulation to clones (68H and 168) isolated from the primary tumor.[6,20] The experiments were done in vivo using s.c. tumors growing in syngeneic mice. The 4.10 tumor was, in general, less sensitive to cytoxan, methotrexate, or fluorouracil compared to 168 and especially 68H.[6] Differential responses to drugs between primary murine tumors and their secondary neoplasms have been described by Fugmann et al.[21] B16 clones with metastatic potential also have an altered response to retinoic acid compared to parental lines.[22] Differential chemosensitivities between primary and metastatic Lewis lung carcinomas have also been demonstrated.[23]

Tsuruo and Fidler performed drug studies on cells isolated from primary tumors and from their metastases.[15] Four tumor models were investigated; the B16 and K-1735 murine melanomas, the OV-2237 mouse fibrosarcoma, and the A-375 human melanoma growing in the nude mouse. Using drugs employed clinically against melanoma, the authors demonstrated that (1) metastases respond differently to antineoplastic agents compared to their parent tumors and (2) distinct metastases from one primary neoplasm or parental cell line also have different sensitivities to drugs compared to one another.[15]

There have been several investigations that have utilized primary and metastatic tumors obtained from patients in order to assess their responsiveness to antineoplastic agents. These studies are of special interest because they evaluate drug sensitivities of primary human neoplasms and their metastases immediately upon removal from patients and thus provide information relevant to the design of adjuvant therapy protocols. Siracky obtained sets of primary and secondary tumor tissues from patients with colon or ovarian carcinoma, and demonstrated that metastases differed in their responses to drugs compared to their primary tumors.[24] Biorklund et al. reported similar findings with different samples of non-Hodgkin's lymphoma from a single patient.[25]

All of the above results strongly indicate that if the clonogenic assay (stem cell assay), developed by Salmon et al.,[26] were applied to fresh human primary *and* secondary tumors from individual patients, differences between primaries and metastases would be demonstrated. Schlag and Schreml did precisely this study, and obtained the predicted result. When a series of disseminated tumors from patients were compared in the stem cell assay with their respective primary neoplasms, a discordance in chemosensitivities was observed.[27] Clearly, a metastasis from a patient can respond dissimilarly to drugs compared to its primary tumor.

This approach has been extended very elegantly by Von Hoff and co-workers.[94,95] These investigators compared chemosensitivites of (1) different regions of each of 13 primary human tumors, (2) a metastasis and primary tumor from each of 29 patients, and (3) two metastases from each of 45 patients. Percent survival differences were greater between a primary neoplasm and its metastasis than between two regions of a primary tumor, confirming the results of Schlag and Schreml. However, among the three groups the greatest differences in survival to drugs were observed between two metastases from a given patient, underscoring the impact of interlesional heterogeneity on adjuvant therapy. This important study suggests the following. First, care must be exercised in employing the stem cell drug sensitivity assay, as results obtained with a patient's primary tumor will not necessarily (or even usually) predict for the "right" drug to be used against all his metastases. This presents limitations to the use of the clonogenic sensitivity assay. Second, interlesional heterogeneity in individual patients for response to treatment modalities is a major obstacle in the complete eradication of disseminated disease in these patients. The clinician is faced with multiple lesions, each with its unique treatment response phenotype. This situation clearly provides one important reason why we have often been unable to cure patients with metastatic disease.

Tanigawa et al. have also recently examined fresh human tumors and their metastases for drug sensitivities in the clonogenic assay. These workers found considerable differences in chemosensitivities between two parts of the same tumor or between a primary tumor and its metastases, for a series of solid neoplasms.[28] This study and those conducted by Schlag and Schreml and by Von Hoff and co-workers illustrate that tissue samples from metastases rather than from primary tumors should be used in the clonogenic assay if it is to have any good probability of predicting responses of patients with solid tumors to anticancer drugs administered in an adjuvant setting.

IV. CLINICAL STUDIES

All the data presented above were obtained in the laboratory whether murine or human tumor lines or fresh tumor specimens were studied. Are there clinical data that bear on this question of differential responses of primary tumors and metastases? Slack and Bross have examined this problem from a different perspective.[29] In a report that must be read to be fully appreciated, they reviewed the results of screening trials of the Eastern Clinical Drug Evaluation Program for 1961 to 1965; response and location data from six primary sites of origin and six metastatic sites were analyzed. There was a total of 1687 lesions included in this assessment of chemotherapy trails. This analysis was published in 1975, before the concept of tumor heterogeneity had emerged, and the results anticipated all that an appreciation of heterogeneity was to teach us.[29] Once again, evidence from clinical work was available that foreshadowed the results of laboratory studies that were designed for their "relevance" to future clinical studies.

The analysis of Slack and Bross revealed a significant amount of variation in response to drug treatment associated with metastatic sites, whereas there was a nonsignificant amount of variation associated with sites of primary origin. The main response of a metastatic tumor depended much more on its location than on the site of the primary tumor from which it disseminated. In general, superficial secondary lesions such as lymph nodes tended to respond whereas deeper lesions (in skeletal or respiratory systems) tended not to respond. Although primary vs. metastases data for individual patients were not analyzed per se, the very large number of lesions examined retrospectively allows for an interpretation entirely consistent with the tenets of heterogeneity for responses of secondary tumors vs. primary neoplasms in individual patients. The authors[29] themselves conclude. "Furthermore, it was apparent that the metastatic lesions that do respond do not give a reliable indication of drug action against parent tumor types." While we might be tempted to state this the other way around today, the message was clearly evident from clinical studies 2 years before Fidler and Kripke reported metastatic heterogeneity in the B16 mouse melanoma. The clinical data were only waiting for the correct insightful biological interpretation.

V. APPROACHES TO THE THERAPY OF METASTATIC DISEASE

Research on the sensitivity of heterogeneous tumors to treatment modalities (primarily drugs) has focused on three types of comparisons. The first comparison is between subpopulations from the same primary neoplasm. This includes studies done on murine tumor clones, subpopulations isolated from established lines of human tumors, and different portions, zones or clones from freshly obtained human primary cancers. As might be expected from our current appreciation of intraneoplastic diversity, differential sensitivities were demonstrated in these various situations (see Chapter 8).

The next level has involved comparing a metastasis with its primary tumor. As reviewed earlier in this chapter, heterogeneity for response to drugs has been shown for surgical specimens of a primary human tumor and a metastasis of that neoplasm, for cells from spontaneous metastases of murine tumors compared to the primary, and for metastatic human melanoma cells in nude mice vs. the parental melanoma line. The approach has been taken to the most clinically relevant level when several metastases in the same mouse, or patient, have been intercompared for their sensitivities to antineoplastic agents. This examination of interlesional heterogeneity for treatment response has provided a window to the problem clinicians face with patients who present with disseminated micro- or macrometastases. The differences demonstrated in such studies underscore the problems in designing efficacious adjuvant therapy protocols and provide a very convincing explanation for our failure to treat successfully the majority of patients with disseminated disease.

The answer to the challenge presented by tumor heterogeneity, particularly interlesional heterogeneity, was perceived (albeit empirically) already several years ago to be multimodality therapy. Distinct subpopulations, often located at several different anatomical sites within the patient, have significantly different responses to various modalities. Moreover, work from our laboratory with the DLD-1 human colon cancer model has demonstrated that the responses of the clone A and clone D subpopulations to various drugs, ionizing radiation, hyperthermia, or differentiation-inducing agents vary with the modality.[18,30-32] Thus, hierarchies of sensitivities exist, whose rankings cannot be predicted.[31] In other words, a clone exquisitely sensitive to radiation might be refractory to drug x, etc., and one cannot *a priori* assign a drug, or radiation as a treatment for any clone based on its known sensitivity or resistance to some other drug. It is clear that multiple clones will require a number of carefully selected agents, and investigators have accepted in principle the statement that heterogeneous tumors must be treated with a heterogeneous collection of *modalities*.[33,34] But how are we to choose the proper agents?

At first glance, the design of some rational combination therapy for metastatic tumors appears to be the proverbial search for the needle in the haystack. A patient with disseminated disease might have 5, 10, 20, or more micro- and macrometastases. Each of these would seem to require the "right" drug for that clone. Furthermore, that series of choices is predicated on the assumption (most likely not true) that each of these several metastases is itself *homogeneous.* If one superimposes the spector of intralesional heterogeneity on the picture of interlesional heterogeneity in the same individual, the situation appears hopeless. One could speculate that the clinician might require 20 or even 50 agents to eradicate all the secondary tumors in a patient. Such a protocol can exist only in our imagination.

There is an approach to the treatment of disseminated disease that looks at the problem from the opposite perspective. What if one were to focus on common features in metastases rather than to continue the frustrating exercise of documenting differences.[35-37] Even though so many phenotypic variations occur in the natural history of a malignancy, might there not be one or a few phenotypic traits that are expressed in a homogeneous fashion by all metastases in a patient, or in a tumor type, or even in all solid tumors? If common denominators exist among metastases, then these commalities will provide us with a handle for the rational design of diagnostic procedures and multimodality therapy for disseminated cancer (Figure 1).

COMMONALITIES SHARED AMONG METASTASES

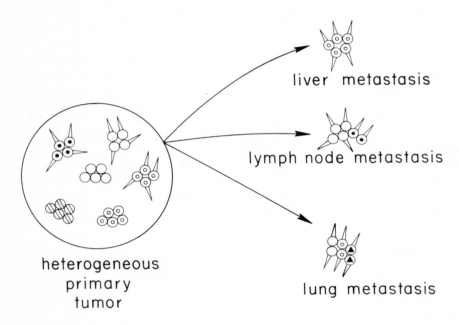

FIGURE 1. Schematic drawing of a common feature shared among all cells in all metastases from a primary tumor in a given host. This commonality may or may not be shared by all nonmetastatic cells in the primary tumor. Where intralesional heterogeneity develops during the evolution of a metastasis (as in the lung metastasis), the new variant clone also expresses the common property. If the shared feature is a surface receptor detectable by antibody or hormone binding, a diagnostic test for identification of occult secondary disease can be envisioned. If the commonality is functional, and can be interdicted by drug treatment for example, a therapy effective against all the cells in the metastatic lesions can be designed. See text for details.

VI. COMMONALITIES IN METASTASES

Is there any evidence that common denominators exist among metastases? There are data that indicate that several types of commonalities might be identified.[35] Each of these common features is amenable, in principle, to attack by an antineoplastic agent. Of course this line of reasoning is still speculative, but sufficient information is now available to suggest that the rational selection of drugs to attack these common properties is a possibility. The value of this approach must, of course, be tested by experimentation. The following list of putative commonalities is based on good experimental data.

A. Ploidy

Twenty-five years ago, Rabotti studied the DNA content of cells in metastases from five human breast tumors, and a nasopharyngeal carcinoma compared to their respective primary neoplasms.[38] The study demonstrated that cells in metastases had higher chromosome numbers than did cells in the corresponding primary cancers. These findings with tumor materials obtained directly from patients are supported by work from our laboratory with cell lines established from human solid tumors (Table 1). Cells from the HCT-15 and DLD-2 lines, established from primary colon carcinomas, are diploid.[17,39] The great majority of cells (clone D) from the heterogeneous DLD-1 pri-

Table 1
PLOIDY OF HUMAN CANCER CELL LINES

Cell line	Tumor type	Location of tumor	Modal chromosome number
HCT-15	Colon carcinoma	Primary colon	45
DLD-2	Colon carcinoma	Primary colon	47
DLD-1	Colon carcinoma	Primary colon	46, 76
Clone D	Colon carcinoma (DLD-1)	Primary colon	46
Clone A	Colon carcinoma (DLD-1)	Primary colon	76
OM-1	Colon carcinoma	Omental metastasis	61
HOT-3	Colon carcinoma	Ovarian metastasis	64
RWP-1	Pancreatic carcinoma	Liver metastasis	64
RWP-2	Pancreatic carcinoma	Liver metastasis	62
H-Mel-3	Melanoma	Superficial node (metastasis)	79

Note: Ploidy of human tumor cell lines established in author's laboratory from primary or metastatic solid tumors in patients. See text for details.

Adapted from Dexter.[35]

mary colon cancer are diploid. The other subpopulation from DLD-1, clone A, is hyperploid with modal chromosome number of 76.[18] Interestingly, clone A is the most poorly differentiated of our lines from primary colorectal cancers and has also been reported to have a higher incidence of metastasis in nude mice than does clone D.[19]

Our results from five cell lines established from patient metastases from three distinct tumor types (colon and pancreatic carcinoma and melanoma) involving four colonization sites show that every line has a hyperploid karyotype, with modal chromosome numbers ranging from 46 to 79 (Table 1).[36,40,41] Our results and those of Rabotti suggest that this increased ploidy demonstrated in human secondary tumors might provide a target for therapy. Agents that crosslink or intercalate with DNA could be examined for effectiveness against metastatic cells.

Any preferential activity against hyperploid cells should provide a basis for a favorable therapeutic index since normal host cells are diploid. A decreased effectiveness against a diploid primary tumor will not pose a problem in most cases since the primary neoplasm is often removed by surgery.

B. Cytoskeleton

The deformability of metastatic cells has been discussed in Chapter 4. The disseminating cell has the ability to squeeze through openings into vessels, round up in the circulation, and finally compress itself out of the vascular bed into the colonization site. Such deformability or plasticity may be accompanied by a locomotor activity.[42-46] All these properties depend on the metastatic cells possessing a flexible cytoskeletal system. This "collapsible" cytoskeleton could prove to be one of the most exploitable commonalities shared by cells in metastases because drugs are available that affect the microtubular/microfilamentous system of cells. Several drugs with this property include vincristine, vinblastine, VP-16, and VM-26, which are plant-derived compounds that are currently used clinically against cancer. Perhaps the inclusion of an agent such as a vinca alkaloid in an adjuvant chemotherapy protocol will provide a selected attack against a sensitive component of metastatic compared to normal cells.

Data from our laboratory lend some support to this hypothesis. Metastatic cells isolated from the C6 rat glioma were more sensitive in every case to vincristine, vinblastine, VP-16, and VM-26 than were parental C6 cells. The variant cells were 33-fold more sensitive to vincristine, an exquisite increase in sensitivity compared to the original cell line.[14]

C. Tumor Angiogenesis Factor

One property of metastatic cells that is absolutely essential for their survival in a distant organ is their ability to elicit a vascular supply from the host. Without such a microvascularization, the incipient tumor nidus will perish from a lack of nutrition and oxygen. The process whereby tumor cells induce the host to provide a blood supply is called angiogenesis, and tumor cells produce a substance called tumor angiogenesis factor (TAF) which is responsible for such vasculature formation.[47-49] An obvious target for secondary tumors growing in the liver, lungs, etc. would be the vasculature network itself, or the factor responsible for its formation. This is an area that has generated intensive interest for over 10 years. Certainly TAF and the vasculature of the secondary tumors is one of the best examples of a critical property shared by metastatic cells that should be vulnerable to chemotherapeutic interdiction.

D. Tumoricidal Macrophages

There is good evidence from experimental studies that activated macrophages can selectively kill tumor cells. Striking results have been obtained that demonstrate that activated macrophages can absolutely discriminate between normal and neoplastic cells, including metastases and temperature-sensitive transformed cells.[50-54] Such findings have led Hart and Fidler to conclude that this selectivity may constitute an absolute difference between normal cells and all tumor cells (especially those in metastases) and thus provide the basis for a tumor-specific therapy of disseminated or inoperable neoplastic disease.[55] Although this hypothesis ultimately must be tested in clinical trials, the sensitivity of cancer cells to macrophages clearly defines an important common denominator among cells from primary and secondary tumors.

E. Cancer Cell Differentiation

There is currently considerable interest in the strategy of inducing differentiation, or a benign state, in cancer cells.[56-58] Such a therapy would not have as its objective the complete killing of all tumor cells in a patient, but rather the stimulation of maturational events in these cells. The differentiated cells might no longer display the malignant phenotype, and they might also be more sensitive to existing modes of therapy. The evidence supporting such an approach has been primarily derived from experimental data demonstrating that various chemicals can induce terminal differentiation in cultured murine and human leukemia cells. Studies with the mouse Friend erythroleukemia and M-1 myelocytic leukemia lines and with HL-60 human promyelocytic leukemia cells have shown that polar solvents, retinoids, and sodium butyrate among other agents are good inducers of maturation in these cells.[59-68] Other investigators including our group have extended these findings to a variety of cultured mouse and human solid tumor cells including murine neuroblastoma, teratoma, and melanoma, and human lung and colon carcinoma, and melanoma.[17,69-76]

The broad spectrum sensitivity of tumor cells to differentiation-inducing chemicals suggests that such responsiveness defines a common feature of neoplastic cells. This stimulation of maturational events in cancer cells could also be exploited in another way. We have reported that the polar solvent N,N-dimethylformamide (DMF) sensitizes cultured mouse mammary and human colon cancer cells to ionizing radiation.[77,78] Moreover, the activities of several purine metabolizing enzymes are modulated by DMF, suggesting that activities of conventional cytotoxic purine analogs might be enhanced by polar solvents.[79] We have recently reported that the effects of *cis*-platinum and mitomycin C on cultured human colon cancer cells are enhanced by pretreating the cells with differentiating concentrations of DMF or its metabolite N-methylformamide (NMF).[80] Langdon et al. have demonstrated that NMF and cyclophosphamide in combination against the M5076 murine tumor were more effective than either agent alone.[81] If tumors in general can be shown to be sensitized to drugs or radiation by exposure to nontoxic doses of differentiating agents, this will define yet another common feature of cancer cells, including those in metastases, that may be exploitable in clinical situations.[56,82]

It is interesting that the possibility that tumor cells could be induced to differentiate to a benign phenotype was anticipated from clinical findings with pediatric neuroblastoma dating back more than 50 years. There have been several documented cases where the neuroblastoma had over a period of several years spontaneously differentiated into benign ganglio-neuroma or mature ganglion sheath and capsular cells; the patients were cured in these instances.[83-87] It is not expected that the great majority of human cancers will differentiate in a spontaneous manner to benign tissues, but the evidence now available provides hope that a variety of tumor types including secondary neoplasms will mature upon stimulation with appropriate differentiating agents. These better differentiated malignancies may behave as benign growths, and they also may be more easily eradicated with conventional therapies.

F. Tumor Cell Kinetics

Studies of kinetic properties of human primary and secondary neoplasms have suggested yet another characteristic that may be shared by metastases. Several investigators have demonstrated that metastases have shorter doubling times compared to primary human tumors; these findings were recently reviewed by Tubiana.[88] Results from Cutler et al.[89] for breast carcinoma and from Breur (presented at the conference on Biological Behavior of Tumors and Chronobiology of Tumors, 1976) for osteosarcoma indicate that metastases appearing early have a shorter doubling time. Charbit et al. reported that the doubling times of metastases are generally shorter compared to those of primary human tumors.[90] These findings suggest that efforts should be made to identify agents that would be effective in an adjuvant setting against small disseminated cancers with a high growth rate or growth fraction. Schabel has discussed the kinetic properties of primary tumors and micrometastases, and how their kinetic behavior will influence drug selection, in his excellent report on this subject in 1975.[91]

G. Oncogenes and their Products

Another commonality has been suggested by recent molecular biology studies that have identified genes in cancer cells that may be responsible for malignancy, called oncogenes.[92] Two different types of investigations over the past several years have provided evidence that oncogenes are important in carcinogenesis and in normal cellular function as well. The first evidence came from work on the retroviruses, especially the Rous sarcoma virus (RSV).[93] A segment of the RSV genome, termed *src*, has been shown to transform infected cells.[94] A cellular homolog of the *src* gene, named *c-src* was subsequently found in normal chicken cells[95] and in other vertebrate cell types;[96] viral and cellular *src* genes code for a similar enzyme product.[97] Anywhere from one to several dozen viral oncogenes with cellular (*c-onc*) counterparts have now been identified and characterized.[92,98]

The second line of experimentation that has led to the identification of oncogenes has involved isolating DNA from cancer cells and transfecting fibroblast 3T3 cells with this DNA to produce malignantly transformed cells.[99,100] Shih and Weinberg demonstrated that a single gene from a human bladder carcinoma cell line could transfer 3T3 cells.[101] This *c-onc* gene was shown to be homologous to the Harvey sarcoma virus transforming gene. Work from several laboratories in recent years has identified a number of *c-onc* genes from tumors or tumor cell lines.[102-104]

Evidence is now accruing that indicates oncogenes are either normal *c-onc* genes that are "overexpressed" in cancer cells, producing too much of some critical product,[105] or they are mutated *c-onc* genes that produce an altered product.[106] The nature of the oncogene products is currently under intensive study. These include kinases,[107] at least one growth factor (platelet-derived growth factor, PDGF),[108] and at least one growth factor receptor (epidermal growth factor receptor).[109] These findings have led to the hypothesis of an oncogene cascade, with *c-onc* genes coding for growth factors and their receptors, whose interactions stimulate other *c-onc* genes involved in cell proliferation. For example, PDGF coded by *c-sis,* stimulates *c-myc,* which in turn is believed to regulate cell growth.[110]

The heterogeneous expression of oncogenes in tumor subpopulations has been discussed in Chapter 3. An exciting area of future research will be to determine whether oncogene products exist within given tumor types which will be shared by all the cells in a patient's metastases, or even by all cells in disseminated tumors in patients with the same type of cancer. The identification of an oncogene product essential for the continued growth of colon cancer metastases, for example, would quickly lead to strategies for therapeutic intervention based on eliminating or neutralizing the crucial oncogene-derived constituent. It may turn out that the crack in the armor of the heterogeneous tumor will be found in the very oncogene(s) that were responsible for the initiation and progressive growth of the neoplasm.

VII. COMMONALITIES FROM ANOTHER PERSPECTIVE

The suggestion that features common to resistant subpopulations in human tumors, particularly as represented in metastatic disease, should be identified in order to provide a rational basis for adjuvant therapy has also been made recently by Goldie and Coldman.[111] In the third major development[111-113] of their innovative thinking on the origins of drug (and other) resistance in human cancers, these authors have considered the human tumor as a heterogeneous neoplasm with mutations to drug resistance arising in clonal subpopulations that have not differentiated out of the stem cell compartment. They have concluded that large human tumors will be quite heterogeneous for what we have called the treatment response phenotype. Because extensive resistance can be expected in a patient with advanced (metastatic) cancer due to the existence of multiple and distinct resistant clones, Goldie and Coldman in their *Conclusions* state the following: "A final word may be said about the extent of heterogeneity that these advanced tumor systems will express. It is apparent from the above considerations that this will be extreme from the point of view of agents selected on the basis of their relatively nonspecific cytotoxicity. Agents that are developed by a more specific process, *directed against phenotypic properties shared by resistant cells present within the tumor,* will, to a degree, overcome the problem associated with random change and reassortment of phenotypic characteristics" (our italics).[111]

We believe that the grim picture presented by heterogeneity especially among metastases in individual patients points to a bleak future for treatment of cancer patients as long as the many differences already convincingly documented are emphasized. What is now required is that investigators stop looking at the differences, at the heterogeneity, and instead begin to focus on identifying common denominators shared among tumor (particularly metastatic) cells. These commonalities will be essential for developing effective diagnostic methods[114] (i.e., expression of a common antigen on all metastases in a patient, which is recognized by a selective monoclonal antibody) as well as therapeutic approaches. Only by identifying the Achilles heel of the entire tumor in one patient, and in series of patients, can we hope to obtain the handle necessary for the design of effective adjuvant therapies.

REFERENCES

1. Sugarbaker, E. V., Cancer metastasis: a product of host tumor interactions, *Curr. Probl. Cancer,* 3, 1, 1979.
2. Cancer Patient Survival Report No. 5, Department of Health, Education, and Welfare Publ. No. (NIH), Washington, D. C., 1977, 77.
3. Sugarbaker, P. H., MacDonald, J. S., and Gunderson, L. L., Colorectal cancer, in *Cancer: Principles and Practices of Oncology,* DeVita, V. T., Jr., Hellman, S., and Rosenberg, S. H., Eds., Lippincott, Philadelphia, 1982, 643.
4. Silverberg, E., Cancer statistics, *Cancer,* 30, 23, 1980.
5. Fisher, B., Slack, N. H., and Bross, I. D., Cancer of the breast: size of the neoplasm and prognosis, *Cancer,* 24, 1071, 1969.
6. Heppner, G. H., Dexter, D. L., DeNucci, T., Miller, F. R., and Calabresi, P., Heterogeneity in drug sensitivity among tumor cell subpopulations of a single mammary tumor, *Cancer Res.,* 38, 3758, 1978.
7. Fidler, I. J. and Kripke, M. L., Metastasis results from pre-existing variant cells within a malignant tumor, *Science,* 197, 893, 1977.
8. Poste, G., Doll, J., Hart, I. R., and Fidler, I. J., *In vitro* selection of murine B16 melanoma variants with enhanced tissue-invasive properties, *Cancer Res.,* 40, 1636, 1980.
9. Hart, I. R., The selection and characterization of an invasive variant of the B16 melanoma, *Am. J. Pathol.,* 97, 387, 1979.
10. Reading, C. L., Belloni, P. N., and Nicolson, G. L., Selection and *in vivo* properties of lectin-attachment variants of malignant murine lymphosarcoma cell lines, *J. Natl. Cancer Inst.,* 64, 1241, 1980.
11. Nicolson, G. L., Brunson, K. W., and Fidler, I. J., Specificity of arrest, survival and growth of selected metastatic variant cell lines, *Cancer Res.,* 38, 4105, 1978.
12. Miner, K. M., Kawaguchi, T., Uba, G. W., and Nicolson, G. L., Clonal drift of cell surface, melanogenic, and experimental metastatic properties of *in vivo* - selected brain meninges-colonizing murine B16 melanoma, *Cancer Res.,* 42, 4631, 1982.
13. Talmadge, J. E. and Fidler, I. J., Enchanced metastic potential of tumor cells harvested from spontaneous metastases of heterogeneous murine tumors, *J. Natl. Cancer Inst.,* 69, 975, 1982.
14. Dexter, D. L., Lee, E. S., DeFusco, D. J., Libbey, P. N., Spremulli, E. N., and Calabresi, P., Selection of metastatic variants from heterogeneous tumor cell lines using the chicken chorioallantoic membrane and nude mouse, *Cancer Res.,* 43, 1733, 1983.
15. Tsuruo, T. and Fidler, I. J., Differences in drug sensitivities among tumor cells from parental tumors, selected variants and spontaneous metastases, *Cancer Res.,* 41, 3058, 1981.
16. Hanna, N. and Fidler, I. J., Expression of metastatic potential of allogeneic neoplasms in young nude mice, *Cancer Res.,* 41, 438, 1981.
17. Dexter, D. L., Barbosa, J. A., and Calabresi, P., N-N-Dimethylformamide-induced alteration of cell culture characteristics and loss of tumorigenicity in cultured human colon carcinoma cells, *Cancer Res.,* 39, 1020, 1979.
18. Dexter, D. L., Spremulli, E. N., Fligiel, Z., Barbosa, J. A., Vogel, R., VanVoorhees, A., and Calabresi, P., Heterogeneity of cancer cells from a single human colon carcinoma, *Am. J. Med.,* 71, 949, 1981.
19. Spremulli, E. N., Dexter, D. L., Young, P., Campbell, D., and Calabresi, P., Clonal origin of spontaneous metastases of a human colon carcinoma xenografted in nude mice, *Proc. Am. Assoc. Cancer Res.,* 24, 28, 1983.
20. Dexter, D. L., Kowalski, H. L., Blazar, B. A., Fligiel, Z., Vogel, R., and Heppner, G. H., Heterogeneity of tumor cells from a single mouse mammary tumor, *Cancer Res.,* 38, 3174, 1978.
21. Fugmann, R. A., Anderson, J. L., Stolfi, R. L., and Martin, D. S., Comparison of adjuvant chemotherapeutic activity against primary and metastatic spontaneous murine tumors, *Cancer Res.,* 37, 496, 1977.

22. Lotan, R. and Nicolson, G. L., Heterogeneity in growth inhibition by B-trans-retinoic acid of metastatic B16 melanoma clones and in vivo-selected cell variant lines, *Cancer Res.*, 39, 4767, 1979.
23. Trope, C., Different sensitivity to cytostatic drugs of primary tumor and metastasis of the Lewis carcinoma, *Neoplasma*, 22, 171, 1975.
24. Siracky, J., An approach to the problem of heterogeneity of human tumor cell populations, *Br. J. Cancer*, 32, 78, 1979.
25. Biorklund, A., Hakansson, L., Stenstarn, B., Trope, C., and Ackerman, M., Heterogeneity of non-Hodgkin's lymphomas as regards sensitivity to cytostatic drugs, *Eur. J. Cancer*, 16, 647, 1980.
26. Salmon, S. E., Hamburger, A. W., Soehnlen, B., Durie, B. G. M., Alberts, D. S., and Moon, T. E., Quantitation of differential sensitivity of human-tumor stem cells to anticancer drugs, *N. Engl. J. Med.*, 298, 1321, 1978.
27. Schlag, P. and Schreml, W., Heterogeneity in growth pattern and drug sensitivity of primary tumor and metastases in the human tumor colony-forming assay, *Cancer Res.*, 42, 4086, 1982.
28. Tanigawa, N., Mizuno, Y., Hashimura, T., Honda, K., Satomura, K., Hikasa, Y., Niwa, O., Sugahara, T., Yoshida, O., Kern, D. H., and Morton, D. L., Comparison of drug sensitivity amoung tumor cells within a tumor, between primary tumor and metastases, and between different metastases in the human tumor colony-forming assay, *Cancer Res.*, 44, 2309, 1984.
29. Slack, N. H. and Bross, I. D. J., The influence of site of metastasis on tumour growth and response to chemotherapy, *Br. J. Cancer*, 32, 78, 1975.
30. Leith, J. T., DeWyngaert, J. K., Dexter, D. L., Calabrsi, P., and Glicksman, A. S., Differential sensitivity of three human colon adenocarcinoma lines to hyperthermic cell killing, *J. Natl. Cancer Inst.*, 61, 381, 1982.
31. Leith, J. T., Dexter, D. L., DeWyngaert, J. K., Zeman, E. M., Cha, M. Y., Calabresi, P., and Glicksman, A. S., Differential response to x-irradiation of subpopulations of two heterogeneous human carcinomas in vitro, *Cancer Res.*, 42, 2556, 1982.
32. Hagar, J. C., Gold, D. V., Barbosa, J. A., Fligiel, Z., Miller, F., and Dexter, D. L., N,N-Dimethylformamide-induced modulation of organ- and tumor-associated markers in cultured human colon carcinoma cells, *J. Natl. Cancer Inst.*, 64, 439, 1980.
33. Calabresi, P., Dexter, D. L., and Heppner, G. H., Clinical and pharmacological implications of cancer cell differentiation and heterogeneity, *Biochem. Pharmacol.*, 28, 1933, 1979.
34. Calabresi, P. and Dexter, D. L., Clinical implications of cancer cell heterogeneity, in *Tumor Cell Heterogeneity: Origins and Implications,* 4th Ann. Bristol-Myers Symp. on Cancer Research, Owens, A. H., Jr., Coffey, D. S., and Baylin, S. B., Eds., Academic Press, New York, 1982, 181.
35. Dexter, D. L., Heterogeneity in human colon cancer, in *Tumor Invasion and Metastasis: Biologic and Therapeutic Aspects,* Nicolson, G. L. and Milas, L., Eds., Raven Press, New York, 1984, 265.
36. Spremulli, E. N., Scott, C., Campbell, D. E., Libbey, N. P., Schochat, D., Gold, D. V., and Dexter, D. L., Characterization of two metastatic subpopulations originating from a single human colon carcinoma, *Cancer Res.*, 43, 3828, 1983.
37. Spremulli, E. N. and Dexter, D. L., Human tumor cell heterogeneity and metastasis, *J. Clin. Oncol.*, 1, 496, 1983.
38. Rabotti, G., Ploidy of primary and metastatic human tumours, *Nature (London)*, 183, 1276, 1959.
39. Crabtree, G. W., Dexter, D. L., Stoeckler, J. D., Savarese, T. M., Ghoda, L. Y., Rogler-Brown, T. L., Calabresi, P., and Parks, R. E., Jr., Activities of purine metabolizing enzymes in human colon carcinoma cell lines and xenograft tumors, *Biochem. Pharmacol.*, 30, 793, 1981.
40. Dexter, D. L., Matook, G. M., Meitner, P. A., Bogaars, H. A., Jolly, G. A., Turner, M. D., and Calabresi, P., Establishment and characterization of two human pancreatic cancer cell lines tumorigenic in athymic mice, *Cancer Res.*, 42, 2705, 1982.
41. Spremulli, E., Dexter, D., McCarthy, K., Matook, G., Libbey, P., and Calabresi, P., Characterization of two new human melanoma cell lines, *Clin. Res.*, 30, 423A, 1982.
42. Zeidman, I., The fate of circulating tumor cells. I. Passage of cells through capillaries, *Cancer Res.*, 21, 38, 1961.
43. Sato, H., Khato, J., Sato, T., and Suzuki, M., Deformability and filtrability of tumor cells through "nucleopore" filter, with reference to viability and metastatic spread, in *Cancer Metastasis, Approaches to the Mechanism, Prevention and Treatment,* Stansly, P. G. and Sato, H., Eds., University of Tokyo Press, Tokyo, 1977, 53.
44. Felix, H. and Strauli, P., Different distribution of 100-A filaments in resting and locomotive leukemia cells, *Nature (London)*, 261, 604, 1976.
45. Strauli, P. and Weiss, L., Cell locomotion and tumor penetration. Report on a workshop of the EORTC Cell Surface Project Group, *Eur. J. Cancer*, 5, 1, 1977.
46. Mareel, M. M., Malignant and nonmalignant cells: structural similarities and behavioural differences, *Experientia*, 36, 510, 1980.
47. Folkman, J., Merler, E., Abernathy, C., and Williams, G., Isolation of a tumor factor responsible for angiogenesis, *J. Exp. Med.*, 133, 275, 1971.

48. Folkman, J., Tumor angiogenesis, *Adv. Cancer Res.,* 19, 331, 1974.
49. Folkman, J., Langer, R., Linhardt, R. J., Handenschild, C., and Taylor, S., Angiogenesis inhibition and tumor regression caused by heparin or a heparin fragment in the presence of cortisone, *Science,* 221, 719, 1983.
50. Fidler, I. J., Inhibition of pulmonary metastasis by intravenous injection of specifically activated macrophages, *Cancer Res.,* 34, 1974, 1974.
51. Hibbs, J. B., Jr., Discrimination between neoplastic and nonneoplastic cells in vitro by activated macrophages, *J. Natl. Cancer Inst.,* 53, 1487, 1974.
52. Liotta, L. A., Gattozzi, C., Kleinerman, J., and Saidel, G., Reduction of tumor cell entry into vessels by BCG-activated macrophages, *Br. J. Cancer,* 36, 639, 1977.
53. Fidler, I. J., Robbin, R. O., and Poste, G., In vitro tumoricidal activity of macrophages against virus-transformed lines with temperature-dependent transformed phenotypic characteristics, *Cell Immunol.,* 38, 131, 1978.
54. Fidler, I. J., Therapy of spontaneous metastases by intravenous injection of liposomes containing lymphokines, *Science,* 208, 1469, 1980.
55. Hart, I. R. and Fidler, I. J., The implications of tumor heterogeneity for studies on the biology and therapy of cancer metastasis, *Biochim. Biophys. Acta,* 651, 37, 1981.
56. Dexter, D. L. and Calabresi, P., Cancer cell differentiation, in *Pancreatic Tumors in Children,* Humphrey, G. B. et al., Eds., Martinus Nijhoff, The Hague, 1982, 45.
57. Pierce, G. B., The benign cells of malignant tumors, in, *Developmental Aspects of Carcinogenesis and Immunity,* King, T. J., Ed., Academic Press, New York, 1974, 3.
58. Sachs, L., The differentiation of myeloid leukemia cells: new possibilities for therapy, *Br. J. Haematol.,* 40, 509, 1978.
59. Friend, C., Scher, W., Holland, J. G., and Sato, T., Hemoglobin synthesis in murine virus-induced leukemic cells in vitro: stimulation of erythroid differentiation by dimethylsulfoxide, *Proc. Natl. Acad. Sci. U.S.A.,* 68, 378, 1971.
60. Scher, W., Preisler, H. D., and Friend, C., Hemoglobin synthesis in murine virus-induced leukemic cells in vitro. III. Effects of 5-bromo-2′-deoxyuridine, dimethylformamide and dimethylsulfoxide, *J. Cell Physiol.,* 81, 63, 1973.
61. Tanaka, M., Levy, J., Terada, M., Breslow, R., Rifkind, R. A., and Marks, P. A., Induction of erythroid differentiation in murine virus infected erythroleukemia cells by highly polar compounds, *Proc. Natl. Acad. Sci. Sci. U.S.A.,* 72, 1003, 1975.
62. Reuben, R. C., Wife, R. L., Breslow, R., Rifkind, R. A., and Marks, P. A., A new group of potent inducers of differentiation in murine erythroleukemia cells, *Proc. Natl. Acad. Sci. U.S.A.,* 73, 862, 1976.
63. Leder, A. and Leder, P., Butyric acid, a potent inducer of erythroid differentiation in cultured erythroleukemic cells, *Cell,* 5, 319, 1975.
64. Terada, M., Epner, E., Nudel, V., Salmon, J., Fibach, E., Rifkind, R. A., and Marks, P. A., Induction of murine erythroleukemia differentiation by actinomycin D, *Proc. Natl. Acad. Sci. U.S.A.,* 75, 2795, 1978.
65. Honma, Y., Kasukabe, T., Okabe, J., and Hozumi, M., Prolongation of survival times of mice inoculated with myeloid leukemia cells by inducers of normal differentiation, *Cancer Res.,* 39, 3167, 1979.
66. Collins, S. J., Ruscetti, F. W., Gallagher, R. E., and Gallo, R. C., Terminal differentiation of human promyelocytic leukemia cells induced by dimethyl sulfoxide and other polar compounds, *Proc. Natl. Acad. Sci. U.S.A.,* 75, 2458, 1978.
67. Collins, S. J., Bodner, A., Tinge, R., and Gallo, R. C., Induction of morphological and functional differentiation of human promyelocytic leukemia cells (HL-60) by compounds which induce differentiation of murine leukemia cells, *Int. J. Cancer,* 25, 213, 1980.
68. Fontana, J. H., Colbert, D. H., and Deisseroth, A. B., Identification of a population of bipotent stem cells in the HL-60 human promyelocytic leukemia cell line, *Proc. Natl. Acad. Sci. U.S.A.,* 78, 3863, 1981.
69. Kimhi, Y., Palfrey, C., Spector, I., Barak, Y., and Littauer, U. Z., Maturation of neuroblastoma cells in the presence of dimethylsulfoxide, *Proc. Natl. Acad. Sci. U.S.A.,* 73, 462, 1976.
70. Strickland, S. and Mahdavi, V., The induction of differentiation in teratocarcinoma stem cells by retinoic acid, *Cell,* 15, 393, 1978.
71. Kreider, J. W., Wade, D. R., Rosenthal, M., and Densley, T., Maturation and differentiation of B16 melanoma cells induced by theophylline treatment, *J. Natl. Cancer Inst.,* 54, 1457, 1975.
72. Tralka, T. S. and Rabson, A. S., Cilia formation in cultures of human lung cancer cells treated with dimethyl sulfoxide, *J. Natl. Cancer Inst.,* 57, 1383, 1976.
73. Huberman, E., Heckman, C., and Langenbach, R., Stimulation of differentiated functions in human melanoma cells by tumor-promoting agents and dimethyl sulfoxide, *Cancer Res.,* 39, 2618, 1979.

74. Meyskens, F. L., Jr. and Fuller, B. B., Characterization of the effects of different retinoids on the growth and differentiation of a human melanoma cell line and selected subclones, *Cancer Res.*, 40, 2194, 1980.
75. Lotan, R. and Lotan, D., Stimulation of melanogenesis in a human melanoma cell line by retinoids, *Cancer Res.*, 40, 3345, 1980.
76. Kim, Y. S., Tasao, D., Siddiqui, B., Whitehead, J. S., Arnstein, P., Bennett, J. J., and Hicks, J., Effects of sodium butyrate and dimethylsulfoxide on biochemical properties of human colon cancer cells, *Cancer*, 45, 1185, 1980.
77. Leith, J. T., Brenner, H. J., DeWyngaert, J. K., Dexter, D. L., Calabresi, P., and Glicksman, A. S., Selective modification of the X-ray survival response of two mouse mammary adenocarcinoma sublines by N,N-dimethylformamide, *Int. J. Radiat. Oncol. Biol. Phys.*, 7, 943, 1981.
78. Leith, J. T., Gaskins, L. A., Dexter, D. L., Calabresi, P. and Glicksman, A. S., Alteration of the survival response of two human colon carcinoma subpopulations to x-irradiation by N,N-dimethylformamide, *Cancer Res.*, 42, 30, 1982.
79. Dexter, D. L., Crabtree, G. W., Stoeckler, J. D., Savarese, T. M., Ghoda, L. Y., Rogler-Brown, T. L., Parks, R. E., Jr., and Calabresi, P., N,N-dimethylformamide and sodium butyrate modulation of the activities of purine-metabolizing enzymes in cultured human colon carcinoma cells, *Cancer Res.*, 41, 808, 1981.
80. Dexter, D. L., DeFusco, D. J., McCarthy, K., and Calabresi, P., Polar solvents increase the sensitivity of cultured human colon cancer cells to cis-platinum and mitomycin C, *Proc. Am. Assoc. Cancer Res.*, 24, 267, 1983.
81. Langdon, S. P., Hickman, J. A., Gescher, G., and Stevens, M. F. G., N-methylformamide (NSC 3051): a potential candidate for combination chemotherapy, *Proc. Am. Assoc. Cancer Res.*, 24, 271, 1983.
82. Dexter, D. L., Leith, J. T., Crabtree, G. W., Parks, R. E., Jr., Glicksman, A. S., and Calabresi, P., N,N-Dimethylformamide-induced modulation of responses of tumor cells to conventional anti-cancer treatment modalities, in *Maturation Factors and Cancer,* Moore, M. A. S., Ed., Raven Press, New York, 1982, 105.
83. Cushing, H. and Wolbach, S. B., The transformation of a malignant paravertebral sympathicoblastoma into a benign ganglioneuroma, *Am. J. Pathol.*, 3, 62, 1927.
84. Kissane, J. M. and Ackerman, L. V., Maturation of tumors of the sympathetic nervous system, *J. Fac. Radiol.*, 7, 109, 1955.
85. Fox, F., Davidson, J., and Thomas, L. B., Maturation of sympathicoblastoma into ganglioneuroma, *Cancer*, 12, 108, 1959.
86. Visfeldt, J., Transformation of sympathicoblastoma into ganglioneuroma, *Acta Pathol. Microbiol. Scand.*, 58, 414, 1963.
87. Dyke, P. C. and Mulkey, D. A., Maturation of ganglioneuroblastoma to ganglioneuroma, *Cancer*, 20, 1343, 1967.
88. Tubiana, M., L. H., Gray medal lecture: cell kinetics and radiation oncology, *Int. J. Radiat. Oncol. Biol. Phys.*, 8, 1471, 1982.
89. Cutler, S. J., Asire, J. A., and Taylor, S. G., Classification of patients with disseminated cancer of the breast, *Cancer*, 24, 861, 1969.
90. Charbit, A., Malaise, E. P., and Tubiana, M., Relation between the pathological nature and growth rate of human tumours, *Eur. J. Cancer,* 7, 307, 1971.
91. Schabel, F. M., Jr., Concepts for the systemic treatment of micrometastases, *Cancer,* 35, 15, 1975.
92. Land, H., Parada, H. F., and Weinberg, R. A., Cellular oncogenes and multistep carcinogenesis, *Science,* 222, 771, 1983.
93. Rous, P. J., Transmission of a malignant new growth by means of a cell-free filtrate, *J. Am. Med. Assoc.*, 56, 198, 1911.
94. Duesberg, D. H. and Vogt, P. K., Differences between the ribonucleic acids of transforming and non-transforming avian tumor viruses, *Proc. Natl. Acad. Sci. U.S.A.,* 67, 1673, 1970.
95. Stehelin, D., Varmus, H. E., Bishop, J. M., and Vogt, P. K., DNA related to the transforming gene of avian sarcoma viruses in present in normal avian DNA, *Nature (London),* 260, 170, 1976.
96. Opperman, H., Levinson, A. D., Varmus, H. E., Levinton, L., and Bishop, J. M., Uninfected verterbrate cells contain a protein that is closely related to the product of the avian sarcoma virus transforming gene (SRC), *Proc. Natl. Acad. Sci. U.S.A.,* 76, 1804, 1979.
97. Collette, M. S., Brugge, J. S., and Erickson, R. L., Characterization of a normal avian cell protein related to the avian sarcoma virus transforming gene product, *Cell,* 15, 1363, 1978.
98. Buick, R. N. and Pollak, M. N., Perspectives on clonogenic tumor cells, stem cells, and oncogenes, *Cancer Res.*, 44, 4909, 1984.
99. Shih, C., Shilo, B. F., Goldfarb, M. P., Dannenberg, A., and Weinberg, R. A., Passage of phenotypes of chemically transformed cells via transfection of DNA and chromatin, *Proc. Natl. Acad. Sci. U.S.A.,* 76, 5714, 1979.

100. Krontiris, T. G. and Cooper, G. M., Transforming activity of human tumor DNAs, *Proc. Natl. Acad. Sci. U.S.A.*, 78, 1181, 1981.
101. Shih, C. and Weinberg, R. A., Isolation of a transforming sequence from a human bladder carcinoma cell line, *Cell*, 29, 161, 1982.
102. Little, C. D., Nan, M. M., Carney, D. N., Gazdar, A. F., and Minna, J. D., Amplification and expression of the *c-myc* oncogene in human lung cancer cell lines, *Nature (London)*, 306, 194, 1983.
103. McCoy, M. S., Toole, J. J., Cunningham, J. M., Chang, E. H., Lowy, D. R., and Weinberg, R. A., Characterization of a human colon/lung carcinoma oncogene, *Nature (London)*, 302, 79, 1983.
104. Pulciani, S., Santos, E., Lanver, A. V., Long, L. K., Aaronson, S. A., and Barbacid, M., Oncogenes in solid human tumours, *Nature (London)*, 300, 539, 1982.
105. Chang, E. H., Furth, M. E., Scolnick, E. M., and Lowy, D. R., Tumorigenic transformation of mammalian cells induced by a normal human gene homologous to the oncogene of the Harvey murine sarcoma virus, *Nature (London)*, 297, 479, 1982.
106. Reddy, E. P., Reynolds, R. K., Santos, E., and Barbacid, M., A point mutation is responsible for the acquisition of transforming properties by the T24 human bladder carcinoma oncogene, *Nature (London)*, 300, 149, 1982.
107. Sugimoto, Y., Whitman, M., Cantley, L. C., and Erikson, R. L., Evidence that the Rous sarcoma virus transforming gene product phosphorylates phosphatidylinositol and diacylglycerol, *Proc. Natl. Acad. Sci. U.S.A.*, 81, 2117, 1984.
108. Waterfield, M. D., Scrace, G. T., Whittle, N., Stroobant, P., Johnsson, A., Wasteson, A., Westermark, B., Heldin, C. H., Huang, J. S., and Denel, T. F., Platelet-derived growth factor is structurally related to putative transforming protein p28^{sis} of simian sarcoma virus, *Nature (London)*, 304, 35, 1983.
109. Downward, J., Yarden, Y., Mayes, E., Scrace, G., Totty, N., Stockwell, P., Ullrich, A., Schlessinger, J., and Waterfield, M. D., Close similarity of epidermal growth factor receptor and *V-erb-B* oncogene protein sequences, *Nature (London)*, 307, 521, 1984.
110. Kelly, K., Cochran, B. H., Stiles, C. D., and Leder, P., Cell cycle specific regulation of the *c-myc* gene by lymphocyte mitogens and platelet-derived growth factor, *Cell*, 35, 603, 1983.
111. Goldie, J. H. and Coldman, A. J., Quantitative model for multiple levels of drug resistance in clinical tumors, *Cancer Treat. Rep.*, 67, 923, 1983.
112. Goldie, J. H., Coldman, A. J., and Gudauskas, G. H., Rationale for the use of alternating non-cross-resistant chemotherapy, *Cancer Treat. Rep.*, 66, 439 1982.
113. Goldie, J. H. and Coldman, A. J., A mathematic model for relating the drug sensitivity of tumors to their spontaneous mutation rate, *Cancer Treat. Rep.*, 63, 1727, 1979.
114. Olsson, L., Sorensen, H. R., and Behnke, O., Intratumoral phenotypic diversity of cloned human lung tumor cell lines and consequences for analyses with monoclonal antibodies, *Cancer*, 54, 1757, 1984.

Chapter 10

SUMMARY AND CONCLUSIONS

The purpose of this text is to discuss, in a single volume, areas important to an appreciation of the phenomenon of mammalian tumor heterogeneity. The areas covered are, to some extent, the choices of the authors, but we feel that we have represented necessary background information relevant to an understanding of intraneoplastic diversity, and that we have described the current status of research and understanding of such diversity. It is a tribute to the insights of early investigators that we can acknowledge that intrinsic tumor heterogeneity was described over a hundred years ago by Virchow (Chapter 1). It is also gratifying to describe the current sophisticated understanding of heterogeneous human tumor data from the clinic as presented by Von Hoff and others (Chapter 9). The knowledge that remarkable diversity exists between primary tumors and their metastases, and among different metastases is a vital step forward in our progress towards effective therapy of cancer.

As we have emphasized often in the preceding chapters, "heterogeneity" is a term that encompasses, at first glance, many different concepts. These concepts include: intrinsic cellular heterogeneity; cell kinetic heterogeneity; environmental heterogeneity; and heterogeneity in time (tumor progression or differentiation) and space (the interlesional heterogeneity of metastatic tumors). The investigator must, however, be able to integrate these concepts, as they clearly overlap and interact (Figure 1 of Chapter 5), and it would be naive not to consider potential interactions. These concepts emphasize the dynamic nature of cancer, and point out the serious problems this dynamism presents to the clinician. While the final answer to the most effective clinical treatment of malignancy as a heterogeneous entity is not at present at hand, ideas relevant to the strategy of treatment have been presented in the text (Chapters 8 and 9).

We are fortunate that many investigators have provided us with a solid framework upon which to base further research and clinical treatments. In particular, the concept of tumor progression as described by Foulds, and the concept of tumor evolution as provided by Nowell have been significant contributions to the understanding of tumor heterogeneity. Many investigators have performed research to define the limits of responsivity of neoplastic subpopulations to various types of cytotoxic agents, including drugs, ionizing radiation, hyperthermia, etc. The importance of the metastatic potential of certain tumor subpopulations has been ingeniously described by the studies of Fidler, Poste, and other workers. Other studies have provided a conceptual framework to consider whether or not "clonal interactions" exist within solid tumors. The relationship of tumor heterogeneity to the concept of cancer as a disease of differentiation has been discussed by MacKillop et al., and Goldie and Coldman have included this concept in their recent publication on the modeling of drug resistance in heterogeneous solid tumor evolution.

In summary, investigations of "tumor heterogeneity" are varied and the findings stimulate the basic researcher and the clinician. It is certain that additional elegant studies will be performed to provide more insight into the phenomenon itself. There are many areas that require much additional information before a clear picture emerges. For example, the cell kinetic status of heterogeneous solid tumors has not been investigated. A very basic level of tumor heterogeneity lies in the division of cell populations between proliferating (P) cells and quiescent (Q) cells. As any type of cytotoxic treatment of a solid tumor will likely recruit cells from the Q to the P compartment, the implications of this compartmentalization of heterogeneous tumors are important to define. This will require the use of appropriate experimental models.

Another area of study would be to probe the importance of extrinsic heterogeneity in solid tumors, as illustrated by the existance of oxic and anoxic regions within neoplasms. Again, the impact of therapy upon this level of diversity requires more investigation.

A question that has not been answered is whether the information obtained from rodent model tumor systems is relevant to the human disease. More research using heterogeneous human tumor model systems (e.g., as xenograft tumors in "nude" mice) to answer this question is strongly indicated. In this regard, a correlated question concerns the perception of the "stability" of the heterogeneous system. The studies of Poste and co-workers have indicated that "clonal interactions" may exist and that the cellular population stability of a heterogeneous system is reestablished after perturbation with, for example, a cytotoxic agent. Whether this is also true for heterogeneous human tumor systems has not been shown, and indeed, it is possible that such stability may not exist in such human neoplasms. This presence or absence of stability would then have a strong impact upon the philosophy of treatment of heterogeneous malignancies.

The recent exciting advances in the area of molecular biology of cancer also present new challenges and opportunities to elucidate the molecular mechanisms operating in the generation of intraneoplastic diversity. We have only scratched the surface regarding the role of oncogenes and their products in the phenotypic variation existing within individual neoplasms and their metastases.

Ultimately, it is important to provide understanding of intrinsic and extrinsic heterogeneity specifically in regard to human neoplasms. We require appropriate models, insightful basic investigation, and enlightened application to the clinical treatment of human cancer. We feel that the future is very promising in this respect, and believe that the next decade will be even more productive and exciting than was the last decade in terms of the understanding and manipulation of intraneoplastic diversity.

INDEX

A

Accidental, random spreading, 37—39
4'-(9-Acridinylamino)methanesulphone-*m*-anisidine (*m*-AMSA), 99
ACTH, see Adrenocorticotropic hormone
Actinomycin D, 97
Adenocarcinomas, see also specific types, 14, 61
Adenosine triphosphate (ATP), 73
Adjuvant therapies, see also Multimodality therapy, 127, 130, 131
Adrenocorticotropic hormone (ACTH), 14
Adriamycin, 99
Alkaline phosphatase, 18
Alkaloids, see also specific types
  vinca, 127
Alkylating agents, 81
Altered complement of chromosomes, 27
Alveolar rhabdomyosarcoma, 12, 19
m-AMSA, see 4'-(9-Acridinylamino)methanesulphone-*m*-anisidine
Analytical aspects of tumor heterogeneity, 45—57
Anchorage independence, 13
Androgens, 56, 57
Aneuploid cells, 58—61
Angiogenesis, 37, 128
Antibodies, see also specific types
  monoclonal, 13, 130
Anticancer drugs, see Chemotherapy
Antigenic heterogeneity, see also Immunological heterogeneity, 3—4, 13
Antigenicity, 6
Antigens, see also specific types
  mouse mammary tumor virus (MMTV), 4, 83
Ascitic tumors, 38, 72
  mouse, 99
Assays, see also specific types, 83, 99
  clonogenic, 123, 124
  fluctuation, 38
  immunohistochemical, 1
  immunological, 4
  in vivo metastatic, 85—91
  stem cell drug sensitivity, 124
ATP, see Adenosine triphosphate

B

BCNU, see 1,3-Bis (2-chloroethyl)-1-nitrosourea
Bicarbonate-carbon dioxide system, 73
Bioassays, see also Assays, 81
Biochemical heterogeneity, 17
Biochemical markers, 4—6, 11, 15
1,3-Bis (2-chloroethyl)-1-nitrosourea (BCNU), 99
Bladder tumors, 60, 129
Blood circulation, 36, 37
Bone marrow, 26
Bone tumors (osteosarcoma), 38, 58, 60
Brain tumors, see also Gliomas, 1, 18, 88
  rat, 99
Breast cancer, see Mammary cancer
Bronchial carcinoma, 69
B16 tumors, 38, 40, 102, 110, 122
  metastases of, 25
  mouse, 55, 86, 99, 124
  murine, 6, 24
Butyrate, 29

C

Calcitonin, 14, 17, 29
CAM, see Chorioallantoic membrane
Cancer, see also specific types
  bladder, 129
  bone, 38
  breast, see Mammary carcinoma
  bronchial, 69
  cervical, 69
  colon, see Colon carcinoma
  colorectal, 12, 15—16, 38, 60
  drugs against, see Chemotherapy
  ductal, 12
  endometrial, 18
  gastric, 38
  gastrointestinal, 58
  genitourinary, 58
  liver, 38, 72, 121, 122, 128
  lung, see Lung carcinoma
  mammary, see Mammary carcinoma
  nasopharyngeal, 126
  ovarian, 12, 17—18, 60, 123
  pancreatic, 38, 121
  prostatic, 38, 56, 57
  rectal, 15
  testicular, 12, 18
  thyroid, see Thyroid carcinoma
Capillary physiology, 37
Carbon dioxide-bicarbonate system, 72
Carbonic anhydrase, 72
Carcinomas, see also specific types, 100
  bladder, 129
  breast, see Mammary carcinoma
  bronchial, 69
  cervical, 69
  clear cell thyroid, 12, 19
  colon, see Colon carcinoma
  colorectal, 12, 15—16, 38, 60
  epidermoid, see Epidermoid carcinoma
  infiltrating ductal, 12
  lung, see Lung carcinoma
  mammary, see Mammary carcinoma
  medullary thyroid (MTC), 12, 16—17, 28, 29
  nasopharyngeal, 126
  ovarian, 12, 17—18, 60, 123
  pancreatic, 38, 121

prostatic, 38, 56, 57
rectal, 15
small cell, see Small cell carcinomas
squamous, see Squamous cell carcinomas
thyroid, see Thyroid carcinoma
Cell cycle, 53, 56—57, 73, 74, 85, 105
Cells, see also specific types
   aneuploid, 58—61
   differentiation of, see Differentiation
   diploid, 60—62, 126
   Ehrlich ascites, 72
   end, 26
   endothelial, 37, 46
   epithelial, 46
   estrogen-receptor negative, see Estrogen-receptor negative cells
   estrogen-receptor positive, 4, 11, 12, 30, 60
   fibroblast 3T3, 129
   hyperploid, 27, 40, 62, 127
   hypoploid, 62
   hypoxic, see Hypoxic cells
   intrinsic heterogeneity within, 137
   invasive, 122
   karyotypic diversity among, 4
   kinetics of, 129, 137
      clonal interactions among tumor subpopulations, 79, 85, 87
      quantitative aspects of tumor heterogeneity, 45, 56, 57, 62
   locomotor ability of, 36
   mechanical damage to, 37
   migratory mechanisms of, 85
   mitotic, 85
   mutability of, 58
   natural killer (NK), 37, 122
   oncogenes of, 31
   organelles of, 74
   oxic, 70
   proliferating (P), 54, 55, 137
   proliferation of, 72—74, 130
   quiescent (Q), 54, 55, 73, 137
   renewal of, 26
   separation of, see Separation of cells
   stem, see Stem cells
   transformation of, 74
   transition, 26, 31
   yield of, 46, 49
Centrifugation, see also specific types
   elutriation, 45, 48—49, 73, 91
   equilibrium density, 50
Cervical carcinoma, 69
C6 glioma, 123
   rat, 122, 127
Chemotherapy, see also specific agents; specific types, 3, 13, 19, 105, 137
   laboratory studies of sensitivity to, 121—124
   resistance to, 97
   responses to, 97—99
   sensitivity to, 122—124
Chondrosarcomas, 58
Chorioallantoic membrane (CAM), 122, 123

Chromatid exchange, 99
Chromosomal markers, 4—5, 40
Chromosome numbers, 126, 127
Chromosomes
   abnormalities in, 58
   altered complement of, 27
   breaks in, 40
   differences in, 4
   rearrangements of, 40
   variability in contents of, 62
Cigarette smoking, 13
Circulation
   blood, 36, 37
   lymphatic, 36
cis-Platinum, 29, 113, 128
Clear cell thyroid carcinoma, 12, 19
Clinical evidence of kinetic tumor heterogeneity, 57—62
Clinical significance of tumor heterogeneity, 7
Clinical studies, 124
Clonal interactions among tumor subpopulations, 25, 79—93, 137, 138
   in vitro experimental studies, 83—85
   in vivo experimental studies, 79—82
   in vivo metastatic assay systems, 85—91
   observation of, 79
   tumor-host cell interactions, 91—92
Clonality of metastases, 39—41
Clones, 2, 4, 6, 7, 19, 23, 38, 125
   brain tumors, 18
   instability within, 39
   lung cancer, 13—15
   non-growing, 81
   rapidly growing, 81
   slow-growing, 81
Clonogenic assays, 123, 124
Clonogenicity, 74
Collagenase, 36, 46, 49
Collapsible cytoskeleton, 127
Colon carcinoma, 15—16
   DLD-1, see DLD-1 colon carcinoma
   environmental aspects, 72, 73
   metastasis, 40, 128—130
   Nowell hypothesis and, 28, 29
   therapy, 101, 102, 123, 126, 128—130
Colonization, 15
Colorectal cancer, 12, 15—16, 38, 60
Combined modality therapy, see Multimodality therapy
Commonalities, 114, 126—131
Common targets, see Commonalities
C-onc genes, 30, 31, 129, 130
Consumption of oxygen, 73
Creatinine kinase, 14
Culture morphology, 6
CY, see Cyclophosphamide
Cyclic nucleotides, 72
Cyclophosphamide (CY), 81, 123
Cytometric heterogeneity, 61
Cytoplasm, 49
Cytoskeleton, 37, 74, 127—128

collapsible, 127
   protein of, 55
Cytostatic activity, 83
Cytotoxic activity, 83, 92
Cytotoxic agents, see also specific agents; specific types, 29, 30, 48, 54, 71, 74, 87, 91, 110, 137, 138
   responses to, 114
   sensitivity to, 79, 81
Cytoxan, 123

D

"Decathalon winner" concept, 36
Dedifferentiation, 15, 17, 27—29
Diamine oxidase, 17, 29
Differential repair, 101
Differential sedimentation, 51
Differentiation-inducing compounds, see also specific types, 29, 30, 97, 112, 128
Differentiation model, 15, 26—29, 128—129, 137
Diffusibility of glucose, 73
N,N-Dimethylformamide (DMF), 112, 113, 128, 129
Diploid cells, 60—62, 126
Dissociation, 45—48
DLD-1 colon carcinoma, 16, 46, 54, 71, 97, 100, 122, 125
   response to combined modality treatments, 112, 114
   response to hyperthermia, 105—108
DMF, see N,N-Dimethylformamide
DNA, 31, 97, 101, 102, 110, 126, 129
   quantitative aspects of tumor heterogeneity and, 53—62
   synthesis of, 85
DNAse, 49
L-Dopa decarboxylase, 17, 29
Drugs, see also specific drugs; specific types, 7, 125
   anticancer, see Chemotherapy
   antineoplastic, see Chemotherapy
   chemotherapeutic, see Chemotherapy
   cytotoxic, see Cytotoxic agents
   markers of, 41
   metabolism of, 30
   resistance to, 40, 41, 87, 97, 130, 137
   sensitivity to, 19, 83, 99, 122
Ductal carcinoma, 12
Dynamic selection, 38
Dynamic tumors, 23, 25—29, 45

E

Ecosystems, 25, 26, 79
Ehrlich ascites cells, 72
Elastase, 36
Elutriation centrifugation, 45, 48—49, 73, 91
Embolization, 36

Emigration phenomenon, 35
End cells, 26
Endometrial cancer, 18
Endothelial cells, 37, 46
Enolase, 14
Environmental factors, 2, 11, 69—77, 137
   nutrients, 72—74
   tumor hypoxia, 69—71
   tumor pH, 71—72
Enzymes, see also specific types, 46, 49
   proteolytic, 36
Epidermal growth factor receptor, 130
Epidermoid carcinoma, 61, 62
Epithelial cells, 46
Equilibrium density centrifugation, 50
ER, see Estrogen receptor
Erythroleukemia, 128
Estrogen receptor (ER), 11, 12, 60
Estrogen-receptor (ER) negative cells, 4, 11, 12
   of breast cancer, 30
Estrogen-receptor (ER) positive cells, 4, 11, 12, 30, 60
Experimental models of metastases, 121—122
Extracellular pH, 72
Extrinsic factors, 11, 24, 38

F

FCM, see Flow cytometry
Fibroblasts, 46, 129
Fibrosarcomas
   mouse, 123
   murine, 38, 50, 87
Filter barrier concept of lymphatic tumor spread, 37
Flow cytometry (FCM), 13, 14, 17, 45, 46, 48, 51—56, 58, 73
   lung tumors and, 60, 62
Fluctuation assays, 38
Fluorescence, 54
Fluorouracil (5-FU), 98, 123
Friend erythroleukemia, 128
5-FU, see Fluorouracil

G

γ-ray irradiation, 81
Ganglioneuroblastoma, 16
Ganglioneuroma, 16, 129
Gastric cancer, 38
Gastrointestinal cancer, 58
Genes, see also specific types
   c-onc, 30, 31, 129, 130
   retrovirus transforming, 30
Genetic instability, 23
Genitourinary cancer, 58
GFAP, see Glial fibrillary acidic protein
Glial fibrillary acidic protein (GFAP), 1
Gliobastoma multiforme, 1
Gliomas, see also Brain tumors, 1, 12, 18, 28
   C6, see C6 gliomas

small cell, 122
Gliosarcoma, 1
Glucose, 73
Glycolysis, 72
Granulocytes, 50
Growth, 82
  intrinsic, 79
  kinetics of, 37, 88
  polyclonal, 88
  rates of, 6, 91
Growth factor receptors, see also specific types, 130
  epidermal, 130
Growth factors, see also specific types, 30, 130

H

Hamster lymphoma, 89
Harvey sarcoma virus, 129
Heat sensitivity, 110
Heat shock proteins (HSP), 109
Hepatocarcinoma, see Liver cancer
Histological heterogeneity, 14
HL-60 promyelocytic leukemia, 128
Homogeneity, 2, 39, 40
  in L1210 cells, 3
  in response phenotypes, 112
Hormones, see also specific types, 11, 30, 74
  adrenocorticotropic (ACTH), 14
  influence of, 74
  mammary carcinomas induced by, 4
  peptide, 14
Host effects on tumors, 24
Host-tumor cell interactions, 91—92
HSP, see Heat shock proteins
Hyperploid cells, 27, 40, 62, 127
Hyperthermia, 71, 87, 97, 114, 125, 137
  responses to, 104—110
Hypodiploid cells, 62
Hypoxic cells, 50, 51, 71, 105
  oxygenation of, 70
  sensitizers, 69

I

Immune response, 37, 81
Immunogenicity, 4, 91
Immunological assays, 4
Immunological heterogeneity, see also Antigenic heterogeneity, 3—4, 79
Immunosuppressive effect, 81
Immunohistochemical assays, 1
Indodeoxyuridine, 4
Infiltrating ductal carcinoma, 12
Inflammatory responses, 85
Instability, 24, 79, 86, 87
  within clones, 39
  genetic, 23
Interactions, see also Clonal interactions
  clonal, 25, 79—83, 137, 138

tumor and host cells, 91—92
  tumor subpopulations, 25, 79—93, 137, 138
Intracellular pH, 72
Intraneoplastic diversity within tumors, 11—19
Intrinsic factors, 11, 24, 137
Intrinsic growth properties, 79
Invasion, 6, 36, 39
Invasive cells, 122
Invasive phenotypes, 38
Invasive potential, 6
In vitro experimental studies, 82—85
In vivo experimental studies, 79—82
In vivo metastatic assays, 85—91
Irradiation, see Radiation

K

Karyotypic heterogeneity, 4, 6, 12—14, 16—19
Kidney metastases, 89
Killer cells, 37, 122
Kinases, see also specific types, 130
Kinetic heterogeneity, 13, 45, 51
  clinical evidence of, 57—62
Kinetics, 56, 58
  cell, 56, 57, 62, 79, 85, 87, 129, 137
  growth, 37, 88
  reoxygenation, 71

L

L1210 cells, 2, 24
  homogeneous, 3
Laboratory studies of sensitivity to chemotherapy, 122—124
Lactate, 73
Large cell lung cancer, 13, 15, 61
Laryngeal mucosa, 60
LCP, see Lung colonization potential
Lectin, 4, 122
Leucocytes, 58
Leukemias, see also specific types, 3, 28, 29, 128
  L1210, 24
  mouse, 2
  murine, 3, 24
  myelocytic, 128
Lewis lung carcinoma, 122, 123
  nonimmunogenic, 88
Liver cancer, 38, 72, 121, 122, 128
  rat, 87
Local microenvironmental influences, 69
Locomotor ability of tumor cells, 36
Lung carcinoma, 58, 69
  epidermoid, 61
  histological heterogeneity in, 14
  hypoxia, 69
  intratumor heterogeneity, 11, 12
  in vivo metastatic assay systems, 89
  ionizing radiation and, 100, 101
  large cell, see Large cell lung cancer

Lewis, see Lewis lung carcinoma
  metastasis, 38, 40
  metastatic, 6
  nonimmunogenic, 88
  nonsmall cell (NSCLC), 13—15, 60—62, 114
  small cell (SCLC), 13, 15, 28, 29, 62, 121, 144
  squamous cell, 13, 14
  therapy, 122, 128
Lung colonization potential (LCP), 89
Lymphatic circulation, 36
Lymph nodes, 19, 37, 60—62
Lymphocytes, 37, 46, 50, 91
Lymphomas, see also specific types, 12, 58
  hamster, 89
  mouse ascites, 99
  mouse thymic, 4
  non-Hodgkin's, 19, 123
  thymic, 4
Lymphosarcomas, see also specific types
  murine, 122
  RAW117, 87

M

Macrophages, 37, 46, 49, 50, 85, 91, 92
  tumoricidal, 128
Mammary carcinoma, 5, 11—13, 38, 39, 58, 60
  hormone-induced, 4
  in vivo experimental studies, 81
  ionizing radiation and, 102, 103
  metastasis, 38, 39
  mouse, see Mouse mammary carcinoma
  murine, 2
  nutrients, 74
  rat, 87, 103
  therapy, 121, 126
Markers
  biochemical, 4—6, 11, 15
  chromosomal, 4—5, 40
  drug, 41
Maturational agents, see also specific agents, 29—30
MeCCNU (semustine), 97
Mechanical damage to cells, 37
Medullary thyroid carcinoma (MTC), 12, 16—17, 28, 29
Melanin, 4, 88
Melanomas, see also specific types, 4, 12, 18, 29, 58, 99, 114, 128
  B16, see B16 tumors
  mouse, 84, 88
  murine, 6, 40, 123
Metabolism of drugs, 30
Metastases, see also specific types, 2, 6, 11, 35—41, 90, 99, 138
  of B16 tumors, 25
  clonality of, 39—41
  commonalities in, 126—130
  experimental models of, 121—122
  homogeneous, 39

  in vitro studies, 84
  in vivo assay systems, 86
  kidney, 89
  liver, see Liver cancer
  lung, see Lung carcinoma
  monoclonal, 39
  quantitative aspects of tumor heterogeneity, 58, 60, 62
  pulmonary, see Lung carcinoma
  as random process, 37—39
  as rare events, 37
  as selective process, 37—39
  as stochastic events, 38
  therapy of, 121—131
    clinical studies, 124
    sensitivity to chemotherapeutic agents, 122—124
Metastatic assays, 85—91
Metastatic phenotypes, 38, 39, 86—88
Metastatic potential, 6, 24, 55, 79, 86, 89, 91, 99, 122, 137
  clonality, 39, 40
Methotrexate (MTX), 82, 83, 98, 99, 123
Methylcholanthrene-induced murine sarcoma, 3
N-Methylformamide (NMF), 113, 114, 129
Microenvironmental influence, 45, 69, 74
Microfilament/microtubule organization, 55
Microheterogeneity, 39
Microtubule/microfilament organization, 55
Microvascularization, 128
Migratory mechanisms of cells, 85
Mitomycin C, 97, 114, 128
Mitotic cells, 85
MMTV, see Mouse mammary tumor virus
Monoclonal antibodies, 13, 130
Monoclonal tumors, 23, 24, 39—41
Monokine, 92
Morphology, 18, 88
Mouse fibrosarcoma, 123
Mouse Friend erythroleukemia, 128
Mouse leukemias, 2
Mouse lymphoma
  ascite, 99
  thymic, 4
Mouse mammary carcinoma, 3, 4, 18, 53, 56, 128
  clonal interactions among tumor subpopulations, 79, 81—83, 89, 91
  environmental aspects, 71, 73, 74
  therapy, 97, 102, 112, 121, 123
Mouse mammary tumor virus (MMTV) antigens, 4, 83
Mouse melanoma, 84, 88
  B16, 55, 86, 99, 124
Mouse sarcoma, 99
Mouse tumors, see also specific types, 2
MTC, see1 Medullary thyroid carcinoma
MTX, see Methotrexate
Mucosa, 60
Multicellular spheroids, 51
Multimodality therapy, see also Adjuvant therapies, 110—114, 125

"Multiple bullets" concept, 112
Murine leukemia, 3
  L1210, 24
Murine lymphosarcoma, 122
Murine melanomas, 40, 123
  B16, 6, 24
Murine neoplasms, 38
Murine neuroblastoma, 128
Murine sarcoma, 24, 40
  methylcholanthrene-induced, 3
Murine tumors, see also specific types, 124, 125, 129
  heterogeneity within, 3
  paradox of, 2—3
  phenomenology of heterogeneity of, 3—7
Mutation, 23, 26, 28, 58
  rates of, 24, 25, 70
Myelocytic leukemia, 128

N

Nasopharyngeal carcinoma, 126
Natural killer (NK) cells, 37, 122
Neuroblastoma, 12, 16, 129
  murine, 128
Neuroendocrine differentiation, 15
Neuron-specific enolase, 14
Neutron irradiation, 50, 102
Nitrogen mustards, 99
Nitrogen-oxygen mixtures, 71
Nitrosoureas, 99
NK, see Natural killer
NMF, see N-Methylformamide
NMR, see Nuclear magnetic resonance
Non-growing clones, 81
Non-Hodgkin's lymphoma, 19, 123
Nonimmunogenic Lewis lung carcinomas, 88
Nonsmall cell lung carcinoma (NSCLC), 13—15, 60—62, 114
Nowell hypothesis, 4, 17, 23—31, 137
NSCLC, see Nonsmall cell lung carcinoma
Nuclear magnetic resonance (NMR), 74
Nucleotides, 72
Nutrients, see also specific types, 72—74, 128

O

Oncogenes, 30—31, 129—130, 138
  cascade of, 130
  cellular, 31
  products, of, 30
Oral mucosa, 60
Organs, see also specific organs
  as metastatic target, see Target organs
  selection of, 38
Osteosarcoma (bone tumors), 38, 58, 60
Ovarian carcinoma, 12, 17—18, 60, 123
Oxic cells, 70
Oxygen, 70, 71, 128
  concentration of, 45
  consumption of, 73
  hypoxic cells and, 70
  during irradiation, 69
  nitrogen mixture with, 71
Oxygen tension in neoplasms, 69

P

PA, see Plasminogen activator
Pancreatic cancer, 38, 121
PDGF, see Platelet-derived growth factor
Peptide hormones, 14
pH, 45, 71—72, 105
Pharyngeal mucosa, 60
Phenomenology of murine tumor heterogeneity, 3—7
Phenotypes
  invasive, 38
  metastatic, 38, 39, 86—88
  response, see Response phenotypes
Phenotypic characteristics, 5
Phenotypic diversity, 13
Plasmin, 85
Plasminogen activator (PA), 39, 84, 85
Platelet-derived growth factor (PDGF), 130
Platinum, 29, 113, 128
Ploidy, 12, 18, 28, 40, 57, 58, 126—127
P388 mouse leukemia, 2
Polar solvents, see also specific types, 29, 112, 128
Polyclonal tumors, 23, 39—41, 88
Polypeptides, 14
Progression model, 13, 15, 45, 58, 60, 62, 92, 97, 137
  defined, 3
  environmental aspects of tumor heterogeneity, 69, 74
  metastasis and, 35, 38
  Nowell hypothesis, 24—30
Proliferating (P) cells, 54, 55, 137
Proliferation of cells, 72—74, 130
Proliferative heterogeneity, 61
Pronase, 49
Prostaglandins, 74
Prostatic carcinoma, 38, 56, 57
Proteases, 36, 39, 46, 49
Proteins, see also specific types
  cytoskeletal, 55
  glial fibrillaryacidic (GFAP), 1
  heat shock (HSP), 109
Proteolytic enzymes, 36
PVP, 14
Pyruvate, 73

Q

Quantitative aspects of tumor heterogeneity, 45—64
  analytical aspects, 45—57
  clinical evidence, 57—62

Quasithreshold dose, 101
Quiescent (Q) cells, 54, 55, 73, 137

R

Radiation, see also specific types, 13, 51, 80, 87, 97, 105, 113, 125, 128, 137
   γ-ray, 81
   neutron, 50, 102
   oxygen during, 69
   response to, 100—103
   sensitivity to, 71, 110, 112
Random process of metastasis, 37—39
Rapidly growing clones, 81
Rareness of metastasis, 37
Rat brain tumors, 99, 122, 127
Rat hepatocarcinoma, 87
Rat mammary adenocarcinoma, 87, 103, 108
RAW117 lymphosarcoma, 87
Rectal carcinoma, 15—16
Renewal of cells, 26
Reoxygenation kinetics, 71
Repair, 101, 112
Resistance to drugs, 40, 87, 97, 130, 137
Resistant subpopulations, 130
Response
   to chemotherapy, 97—99
   to combined modalities, 110—116
   to cytotoxic agents, 114
   to hyperthermia, 104—110
   inflammatory, 85
   to radiation, 100—105
   to therapy, 7, 97—117
Response phenotypes, 79
   homogenization of, 112
Retinoblastoma, 12, 19
Retinoic acid, 123
Retinoids, 128
Retroviruses, 129
Retrovirus transforming genes, 30
Rhabdomyosarcoma, 12, 19
RIF-1 mouse sarcoma, 99
RNA, 54, 74
Rous sarcoma virus (RSV), 129
RSV, see Rous sarcoma virus

S

Sarcomas, see also specific types, 3
   mouse mammary, see Mouse mammary carcinoma
   murine, see Murine sarcoma
   soft-tissue, 58
SCE, see Sister chromatid exchange
SCLC, see Small cell lung cancer
Sedimentation, 51
Selection, 28, 37—39
   dynamic, 38
   entrinsic, 38
   organ, 38
   pressures of, 23, 25, 71, 74
Semustine (MeCCNU), 97
Sensitivity
   chemotherapy, 122—124
   cytotoxic agent, 79, 81
   drug, 19, 79, 81, 83, 99, 122
   heat, 110
   maturational agent, 29
   radiation, 71, 110, 112, 114
   X-ray, 114
Separation of cells, see also specific methods, 45—48, 50—51
Sister chromatid exchange (SCE), 99
Slow-growing clones, 81
Small cell carcinomas, 15
   lung, see Small cell lung cancer (SCLC)
Small cell glioma, 122
Small cell lung cancer (SCLC), 13—15, 28, 29, 62, 121, 144
Smoking, 13
Sodium burtyrate, 128
Soft-tissue sarcomas, 58
Solid tumors, see also specific types, 59, 124, 126, 128, 137
   clonal interactions among tumor subpopulations, 80, 81, 91
   metastasis, 38, 39
   Nowell hypothesis and, 24, 26, 28, 30
   pH within, 71
   progression in, 35
Spheroids, 50, 55
   multicellular, 51
Squamous cell carcinoma, 29, 58, 60
   of lung, 13, 14
Stability, 25, 86, 138
Stem cell assays, 123, 124
Stem cell-end cell model, 28, 30
Stem cells, 26—29, 31
Stem-transition-end cell model, 29, 30
Stochastic aspects of metastasis, 38
Suspension, 37
Synergism, 112
Synthesis of DNA, 85

T

TAF, see Tumor angiogenesis factor
Target organs, see also specific organs, 2, 38
Teratocarcinomas, 28, 29
Teratoma, 29, 128
Testicular cancer, 12, 18
Thermotolerance, 105, 106, 109
Thermotolerance ratio (TTR), 107
Thymic lymphomas, 4
Thymidine, 85
Thyroid carcinoma, 29
   clear cell, 12, 19
   medullary (MTC), 12, 16—17, 28, 29
Tissue histology, 6

Transformation of cells, 74
Transition cells, 26, 31
Trypsin, 46, 49, 50
TTR, see Thermotolerance ratio
Tumor angiogenesis factor (TAF), 128
Tumor-host cell interactions, 91—92
Tumoricidal macrophages, 128
Tumor potential, 6
Tumor progression, see Progression model

Vincristine, 99, 123, 127
Viruses, see also specific types
   Harvey sarcoma, 129
   mouse mammary tumor (MMTV), 4
   retro-, 129
   Rous sarcoma (RSV), 129
Vitamin A, 74
VM-26, 127
VP-16, 127

U

Utilization of glucose, 73

V

Variability in chromosome content, 62
Vascularization, 37
Vasculature, 36, 72, 98
Vinblastine, 127
Vinca alkaloid, 127

X

X-irradiation, 29, 40, 50, 71, 102, 112
   lesions induced by, 110
   sensitivity of, 114

Z

Zonal heterogeneity, 11, 39, 46, 60, 72
Zonal microheterogeneity, 39

THE LIBRARY
UNIVERSITY OF CALIFORNIA
San Francisco
(415) 476-2335

THIS BOOK IS DUE ON THE LAST DATE STAMPED BELOW
Books not returned on time are subject to fines according to the Library Lending Code. A renewal may be made on certain materials. For details consult Lending Code.

14 DAY AUG 26 1986 RETURNED AUG 25 1986 **14 DAY** NOV - 8 1988 RETURNED NOV - 8 1988		

UCSF LIBRARY MATERIALS MUST BE RETURNED TO:
THE UCSF LIBRARY
530 Parnassus Ave.
University of California, San Francisco
This book is due on the last date stamped below.
Patrons with overdue items are subject to penalties.
Please refer to the Borrower's Policy for details.
Items may be renewed within five days prior to the due date.
For telephone renewals — call (415) 476-2335
Self renewals -- at any UCSF Library Catalog terminal in the Library, or renew by accessing the UCSF Library Catalog via the Library's web site:
http://www.library.ucsf.edu
All items are subject to recall after 7 days.

28 DAY LOAN

28 DAY JUN 1 5 2000 RETURNED JUN 1 5 2000		